Mathematics of the Discrete Fourier Transform (DFT) with Audio Applications

Second Edition

Julius O. Smith III

Center for Computer Research in Music and Acoustics (CCRMA)
Department of Music, Stanford University, Stanford, California 94305 USA

Mathematics of the Discrete Fourier Transform (DFT)

with Audio Applications

Second Edition

Julius O. Smith III

Second printing, May 2008
W3K Publishing (`http://books.w3k.org`)

ISBN 978-0-9745607-4-8

Printed in the United States of America by
BookSurge Publishing (`http://www.booksurge.com`).

Printed from PostScript™ files generated by LaTeX 2_ε and `dvips` using standard 10-point `book` format, and the packages `vmargin`, `fancyheadings`, `graphicx`, `subfigure`, `makeidx`, `latexsym`, `html`, `multicol`, and `boxedminipage`.

Contents

Preface

The Discrete Fourier Transform (DFT) can be understood as a numerical approximation to the *Fourier transform*. However, the DFT has its own exact Fourier theory, which is the main focus of this book. The DFT is normally encountered in practice as a *Fast Fourier Transform* (FFT)—*i.e.*, a high-speed algorithm for computing the DFT. FFTs are used extensively in a wide range of digital signal processing applications, including spectrum analysis, high-speed convolution (linear filtering), filter banks, signal detection and estimation, system identification, audio compression (*e.g.*, MPEG-II AAC), spectral modeling sound synthesis, and many other applications; some of these will be discussed in Chapter 8.

This book started out as a series of readers for my introductory course[1] in digital audio signal processing that I have given at the Center for Computer Research in Music and Acoustics (CCRMA) since 1984. The course was created primarily for entering Music Ph.D. students in the Computer Based Music Theory program at CCRMA. As a result, the only prerequisite is a good high-school math background, including some calculus exposure.

Chapter Outline

1. "Introduction to the DFT" points out the mathematical elements which will be discussed in this book, all motivated by the DFT.

2. "Introduction to Complex Numbers" is about factoring polynomials, the quadratic formula, the complex plane, and Euler's formula.

3. "Proof of Euler's Identity" derives Euler's identity in detail. This is an important tool for working with complex numbers, and one of the critical elements of the DFT definition we need to understand.

4. "Sinusoids and Exponentials" is devoted to two of the most elementary signals in signal processing—sinusoids and exponentials. Also

[1] `http://ccrma.stanford.edu/CCRMA/Courses/320/`

covered are complex sinusoids, audio decay time (t_{60}), in-phase and quadrature sinusoidal components, analytic signals, positive and negative frequencies, constructive and destructive interference, invariance of sinusoidal frequency in linear time-invariant systems, circular motion as the vector sum of in-phase and quadrature sinusoidal motions, sampled sinusoids, and generating sampled sinusoids from powers of a unit-modulus complex number.

5. "Geometric Signal Theory" provides an introduction to vector spaces, inner products, orthogonality, projection of one signal onto another, norms, and elementary vector space operations. In this setting, the DFT can be regarded as a change of coordinates from one basis set (shifted impulses) to another (sinusoids at different frequencies).

6. "The DFT Derived" derives the DFT as a projection of a length N signal $x(\cdot)$ onto the set of N sampled complex sinusoids generated by the Nth roots of unity.

7. "Fourier Theorems for the DFT" derives basic Fourier symmetry relations, the shift theorem, convolution theorem, correlation theorem, power theorem, and theorems pertaining to interpolation and downsampling.

8. "Example Applications of the DFT" illustrates practical FFT analysis in Matlab® and Octave (an open-source matlab) through a series of examples. The various Fourier theorems of the preceding chapter provide a "thinking vocabulary" for understanding these applications.

Elementary and supporting information is provided in a series of appendices. Topics include introductions to sampling theory, Taylor series expansions, logarithms, decibels, digital audio number systems, matrices, round-off noise, Fourier series, and continuous-time Fourier theorems, such as the similarity and differentiation theorems. As a segue to computer-based approaches, a well used Fast Fourier Transform (FFT) algorithm is derived. Finally, various software examples in the Matlab (or Octave) programming language are presented.

This book is first in a series of four texts used in my music/audio signal processing course sequence at CCRMA. Three of the four books (including most of this one) are available on the Web[2] as of this writing.

[2] `http://ccrma.stanford.edu/~jos/pubs.html`

Acknowledgments

Thanks to my graduate students and research colleagues Sean Bratnober, Mark Cartwright, Humane Chan, Rob Hamilton, Miriam Kolar, Randal Leistikow, Sandy Lin, Gautham Mysore, Juhan Nam, Jeonghun Noh, Brook Reeder, Bill Schottstaedt, Sook Young Won, Vivian Woo, and Matt Wright for helpful errata reporting and other suggestions based on earlier precursors to this book. Special thanks are due to Miller Puckette for some especially useful contributions. Thanks also to netizens Greg Allen, Adi Biton, John Brenneise, David Holman, Alexander Kraus, Nima Maftoon, Niels Moseley, Hirak Parikh, and Eric Woudenberg for reporting errata in the draft versions on the Web. Additionally, email discussions with Steven Johnson (co-author of the well known FFTW software package) contributed greatly to the appendix on FFT algorithms.

Many thanks to my 1997 teaching assistant Craig Stuart Sapp for writing the Mathematica magic for Figures 4.9 (adapted for the cover), 4.17, 4.18, 4.19, 7.5, 7.9, 7.11, and 7.12.

Thanks also to Prof. C. Sidney Burrus of Rice University for teaching me and others the fundamentals of digital signal processing, including the DFT.

Finally, thanks to my wife Carol for her encouragement and support, and for putting up with all those "working Saturdays", and to my son Harrison, age 9, for his many inspirations, both musical and mathematical, among others.

Julius Smith
April, 2007

Errata

While I have tried, with help from my students and interested colleagues, to compose a correct manuscript, it is inevitable that "suboptimalities" will be discovered over time. The Web page `http://books.w3k.org/mdft2/` will list any known errata, clarifications, or other information pertaining to this book.

Chapter 1

Introduction to the DFT

This chapter introduces the Discrete Fourier Transform (DFT) and points out the mathematical elements that will be explicated in this book. To find motivation for a detailed study of the DFT, the reader might first peruse Chapter 8 to get a feeling for some of the many practical applications of the DFT. (See also the preface on page xi.)

Before we get started on the DFT, let's look for a moment at the *Fourier transform* (FT) and explain why we are not talking about it instead. The Fourier transform of a continuous-time signal $x(t)$ may be defined as

$$X(\omega) = \int_{-\infty}^{\infty} x(t)e^{-j\omega t}dt, \qquad \omega \in (-\infty, \infty).$$

Thus, right off the bat, we need *calculus*. The DFT, on the other hand, replaces the infinite integral with a finite sum:

$$X(\omega_k) \stackrel{\Delta}{=} \sum_{n=0}^{N-1} x(t_n)e^{-j\omega_k t_n}, \qquad k = 0, 1, 2, \ldots, N-1,$$

where the various quantities in this formula are defined on the next page. Calculus is not needed to define the DFT (or its inverse, as we will see), and with finite summation limits, we cannot encounter difficulties with infinities (provided $x(t_n)$ is finite, which is always true in practice). Moreover, in the field of digital signal processing, signals and spectra are processed only in *sampled* form, so that the DFT is what we really need anyway (implemented using an FFT when possible). In summary, the DFT is *simpler mathematically*, and *more relevant computationally* than the Fourier transform. At the same time, the basic concepts are the same. Therefore, we begin with the DFT, and address FT-specific results in the appendices.

1.1 DFT Definition

The *Discrete Fourier Transform (DFT)* of a signal x may be defined by

$$X(\omega_k) \isdef \sum_{n=0}^{N-1} x(t_n)e^{-j\omega_k t_n}, \qquad k=0,1,2,\ldots,N-1,$$

where '$\isdef$' means "is defined as" or "equals by definition", and

$$\begin{aligned}
\sum_{n=0}^{N-1} f(n) &\isdef f(0)+f(1)+\cdots+f(N-1)\\
x(t_n) &\isdef \text{input signal } \textit{amplitude} \text{ (real or complex) at time } t_n \text{ (sec)}\\
t_n &\isdef nT = n\text{th sampling instant (sec), } n \text{ an integer} \geq 0\\
T &\isdef \text{sampling interval (sec)}\\
X(\omega_k) &\isdef \textit{spectrum} \text{ of } x \text{ (complex valued), at frequency } \omega_k\\
\omega_k &\isdef k\Omega = k\text{th frequency sample (radians per second)}\\
\Omega &\isdef \frac{2\pi}{NT} = \text{radian-frequency sampling interval (rad/sec)}\\
f_s &\isdef 1/T = \textit{sampling rate} \text{ (samples/sec, or Hertz (Hz))}\\
N &= \text{number of time samples = no. frequency samples (integer).}
\end{aligned}$$

The sampling interval T is also called the sampling *period.* For a tutorial on sampling continuous-time signals to obtain non-aliased discrete-time signals, see Appendix D.

When all N signal samples $x(t_n)$ are real, we say $x \in \mathbf{R}^N$. If they may be complex, we write $x \in \mathbf{C}^N$. Finally, $n \in \mathbf{Z}$ means n is any integer.

1.2 Inverse DFT

The *inverse* DFT (the IDFT) is given by

$$x(t_n) = \frac{1}{N}\sum_{k=0}^{N-1} X(\omega_k)e^{j\omega_k t_n}, \qquad n=0,1,2,\ldots,N-1.$$

The inverse DFT is written using '$=$' instead of '$\isdef$' because the result follows from the definition of the DFT, as we will show in Chapter 6.

1.3 Mathematics of the DFT

In the signal processing literature, it is common to write the DFT and its inverse in the more pure form below, obtained by setting $T = 1$ in the previous definition:

$$\begin{aligned} X(k) &\stackrel{\Delta}{=} \sum_{n=0}^{N-1} x(n)e^{-j2\pi nk/N}, \qquad k = 0,1,2,\ldots,N-1 \\ x(n) &= \frac{1}{N}\sum_{k=0}^{N-1} X(k)e^{j2\pi nk/N}, \qquad n = 0,1,2,\ldots,N-1 \end{aligned}$$

where $x(n)$ denotes the input signal at time (sample) n, and $X(k)$ denotes the kth spectral sample. This form is the simplest mathematically, while the previous form is easier to interpret physically.

There are two remaining symbols in the DFT we have not yet defined:

$$\begin{aligned} j &\stackrel{\Delta}{=} \sqrt{-1} \\ e &\stackrel{\Delta}{=} \lim_{n\to\infty}\left(1+\frac{1}{n}\right)^n = 2.71828182845905\ldots \end{aligned}$$

The first, $j = \sqrt{-1}$, is the basis for *complex numbers*.[1] As a result, complex numbers will be the first topic we cover in this book (but only to the extent needed to understand the DFT).

The second, $e = 2.718\ldots$, is a (transcendental) real number defined by the above limit. We will derive e and talk about why it comes up in Chapter 3.

Note that not only do we have complex numbers to contend with, but we have them appearing in exponents, as in

$$s_k(n) \stackrel{\Delta}{=} e^{j2\pi nk/N}.$$

We will systematically develop what we mean by imaginary exponents in order that such mathematical expressions are well defined.

With e, j, and imaginary exponents understood, we can go on to prove *Euler's Identity*:

$$e^{j\theta} = \cos(\theta) + j\sin(\theta)$$

Euler's Identity is the key to understanding the meaning of expressions like

$$s_k(t_n) \stackrel{\Delta}{=} e^{j\omega_k t_n} = \cos(\omega_k t_n) + j\sin(\omega_k t_n).$$

[1]Physicists and mathematicians use i instead of j to denote $\sqrt{-1}$.

We'll see that such an expression defines a *sampled complex sinusoid*, and we'll talk about sinusoids in some detail, particularly from an audio perspective.

Finally, we need to understand what the summation over n is doing in the definition of the DFT. We'll learn that it should be seen as the computation of the *inner product* of the signals x and s_k defined above, so that we may write the DFT, using inner-product notation, as

$$X(k) \stackrel{\Delta}{=} \langle x, s_k \rangle$$

where $s_k(n) \stackrel{\Delta}{=} e^{j2\pi nk/N}$ is the sampled complex sinusoid at (normalized) radian frequency $\omega_k T = 2\pi k/N$, and the inner product operation $\langle \cdot, \cdot \rangle$ is defined by

$$\langle x, y \rangle \stackrel{\Delta}{=} \sum_{n=0}^{N-1} x(n)\overline{y(n)}.$$

We will show that the inner product of x with the kth "basis sinusoid" s_k is a measure of "how much" of s_k is present in x and at "what phase" (since it is a complex number).

After the foregoing, the inverse DFT can be understood as the *sum of projections* of x onto $\{s_k\}_{k=0}^{N-1}$; *i.e.*, we'll show

$$x(n) = \sum_{k=0}^{N-1} \tilde{X}_k s_k(n), \qquad n = 0, 1, 2, \ldots, N-1$$

where

$$\tilde{X}_k \stackrel{\Delta}{=} \frac{X(k)}{N}$$

is the *coefficient of projection* of x onto s_k. Using the notation $x \stackrel{\Delta}{=} x(\cdot)$ to mean the whole signal $x(n)$ for all $n \in [0, N-1]$, the IDFT can be written more simply as

$$x = \sum_k \tilde{X}_k s_k.$$

Note that both the *basis sinusoids* s_k and their coefficients of projection $\tilde{X}_k$ are *complex valued* in general.

Having completely understood the DFT and its inverse mathematically, we go on to proving various *Fourier Theorems*, such as the "shift theorem," the "convolution theorem," and "Parseval's theorem." The Fourier theorems provide a basic thinking vocabulary for working with signals in the time and frequency domains. They can be used to answer questions such as

> "What happens in the frequency domain if I do [operation x] in the time domain?"

Usually a frequency-domain understanding comes closest to a *perceptual* understanding of audio processing.

Finally, we will study a variety of practical spectrum analysis examples, using primarily the `matlab` programming language [64] to analyze and display signals and their spectra.

1.4 DFT Math Outline

In summary, understanding the DFT takes us through the following topics:

- Complex numbers
- Complex exponents
- Why e?
- Euler's identity
- Projecting signals onto signals via the inner product
- The DFT as the coefficient of projection of a signal x onto a sinusoid
- The IDFT as a sum of projections onto sinusoids
- Various Fourier theorems
- Elementary time-frequency pairs
- Practical spectrum analysis in matlab

We will additionally discuss various practical aspects of working with signals and spectra.

Chapter 2

Complex Numbers

This chapter introduces complex numbers, beginning with factoring polynomials, and proceeding on to the complex plane and Euler's identity.

2.1 Factoring a Polynomial

Remember "factoring polynomials"? Consider the second-order polynomial

$$p(x) = x^2 - 5x + 6.$$

It is second-order because the highest power of x is 2 (only non-negative integer powers of x are allowed in this context). The polynomial is also *monic* because its leading coefficient, the coefficient of x^2, is 1. By the fundamental theorem of algebra (discussed further in §2.4), there are exactly two *roots* (or *zeros*) of any second order polynomial. These roots may be real or complex (to be defined). For now, let's assume they are both real and denote them by r_1 and r_2. Then we have $p(r_1) = 0$ and $p(r_2) = 0$, and we can write

$$p(x) = (x - r_1)(x - r_2).$$

This is the *factored form* of the monic polynomial $p(x)$. (For a non-monic polynomial, we may simply divide all coefficients by the first to make it monic, and this doesn't affect the zeros.) Multiplying out the symbolic factored form gives

$$p(x) = (x - r_1)(x - r_2) = x^2 - (r_1 + r_2)x + r_1 r_2.$$

Comparing with the original polynomial, we find we must have

$$\begin{aligned} r_1 + r_2 &= 5 \\ r_1 r_2 &= 6. \end{aligned}$$

This is a system of two equations in two unknowns. Unfortunately, it is a *nonlinear* system of two equations in two unknowns.[1] Nevertheless, because it is so small, the equations are easily solved. In beginning algebra, we did them by hand. However, nowadays we can use a software tool such as Matlab or Octave to solve very large systems of linear equations.

The factored form of this simple example is

$$p(x) = x^2 - 5x + 6 = (x - r_1)(x - r_2) = (x - 2)(x - 3).$$

Note that polynomial factorization rewrites a monic nth-order polynomial as the product of n *first-order* monic polynomials, each of which contributes one zero (root) to the product. This factoring business is often used when working with *digital filters* [65].

2.2 The Quadratic Formula

The general second-order (real) polynomial is

$$p(x) \stackrel{\Delta}{=} ax^2 + bx + c \tag{2.1}$$

where the coefficients a, b, c are any real numbers, and we assume $a \neq 0$ since otherwise it would not be second order. Some experiments plotting $p(x)$ for different values of the coefficients leads one to guess that the curve is always a scaled and translated *parabola*. The canonical parabola centered at $x = x_0$ is given by

$$y(x) = d \cdot (x - x_0)^2 + e \tag{2.2}$$

where the magnitude of d determines the width of the parabola, and e provides an arbitrary vertical offset. If $d > 0$, the parabola has the minimum value e at $x = x_0$; when $d < 0$, the parabola reaches a maximum at $x = x_0$ (also equal to e). If we can find d, e, x_0 in terms of a, b, c for any quadratic polynomial, then we can easily factor the polynomial. This is called *completing the square*. Multiplying out the right-hand side of Eq. (2.2) above, we get

$$y(x) = d(x - x_0)^2 + e = dx^2 - 2dx_0x + dx_0^2 + e. \tag{2.3}$$

[1] "Linear" in this context means that the unknowns are multiplied only by constants—they may not be multiplied by each other or raised to any power other than 1 (*e.g.*, not squared or cubed or raised to the 1/5 power). Linear systems of N equations in N unknowns are very easy to solve compared to *nonlinear* systems of N equations in N unknowns. For example, Matlab and Octave can easily handle them. You learn all about this in a course on *Linear Algebra* which is highly recommended for anyone interested in getting involved with signal processing. Linear algebra also teaches you all about *matrices*, which are introduced only briefly in Appendix H.

Equating coefficients of like powers of x to the general second-order polynomial in Eq. (2.1) gives

$$\begin{aligned} d &= a \\ -2dx_0 &= b \quad \Rightarrow \quad x_0 = -b/(2a) \\ dx_0^2 + e &= c \quad \Rightarrow \quad e = c - b^2/(4a). \end{aligned}$$

Using these answers, any second-order polynomial $p(x) = ax^2 + bx + c$ can be rewritten as a scaled, translated parabola

$$p(x) = a\left(x + \frac{b}{2a}\right)^2 + \left(c - \frac{b^2}{4a}\right).$$

In this form, the roots are easily found by solving $p(x) = 0$ to get

$$\boxed{x = \frac{-b \pm \sqrt{b^2 - 4ac}}{2a}.}$$

This is the general *quadratic formula.* It was obtained by simple algebraic manipulation of the original polynomial. There is only one "catch." What happens when $b^2 - 4ac$ is negative? This introduces the square root of a negative number which we could insist "does not exist." Alternatively, we could invent complex numbers to accommodate it.

2.3 Complex Roots

As a simple example, let $a = 1$, $b = 0$, and $c = 4$, *i.e.*,

$$p(x) = x^2 + 4.$$

As shown in Fig. 2.1, this is a parabola centered at $x = 0$ (where $p(0) = 4$) and reaching upward to positive infinity, never going below 4. It has no real zeros. On the other hand, the quadratic formula says that the "roots" are given formally by $x = \pm\sqrt{-4} = \pm 2\sqrt{-1}$. The square root of any negative number $c < 0$ can be expressed as $\sqrt{|c|}\sqrt{-1}$, so the only new algebraic object is $\sqrt{-1}$. Let's give it a name:

$$\boxed{j \triangleq \sqrt{-1}}$$

Then, formally, the roots of $x^2 + 4$ are $\pm 2j$, and we can formally express the polynomial in terms of its roots as

$$p(x) = (x + 2j)(x - 2j).$$

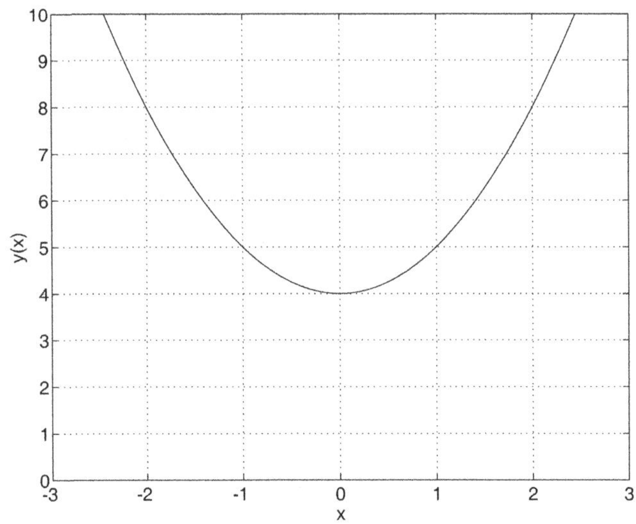

Figure 2.1: An example parabola defined by $p(x) = x^2 + 4$.

We can think of these as "imaginary roots" in the sense that square roots of negative numbers don't really exist, or we can extend the concept of "roots" to allow for *complex numbers*, that is, numbers of the form

$$z = x + jy$$

where x and y are real numbers, and $j^2 \stackrel{\Delta}{=} -1$.

It can be checked that all algebraic operations for real numbers[2] apply equally well to complex numbers. Both real numbers and complex numbers are examples of a mathematical *field*.[3] Fields are *closed* with respect to multiplication and addition, and all the rules of algebra we use in manipulating polynomials with real coefficients (and roots) carry over unchanged to polynomials with complex coefficients and roots. In fact, the rules of algebra become simpler for complex numbers because, as discussed in the next section, we can *always* factor polynomials completely over the field of complex numbers while we cannot do this over the reals (as we saw in the example $p(x) = x^2 + 4$).

[2](multiplication, addition, division, distributivity of multiplication over addition, commutativity of multiplication and addition)

[3]See, *e.g.*, Eric Weisstein's *World of Mathematics* (`http://mathworld.wolfram.com/`) for definitions of any unfamiliar mathematical terms such as a *field* (which is described, for example, at the easily guessed URL `http://mathworld.wolfram.com/Field.html`).

2.4 Fundamental Theorem of Algebra

Every nth-order polynomial possesses exactly n complex roots.

This is a very powerful algebraic tool.[4] It says that given any polynomial

$$\begin{aligned} p(x) &= a_n x^n + a_{n-1}x^{n-1} + a_{n-2}x^{n-2} + \cdots + a_2x^2 + a_1x + a_0 \\ &\triangleq \sum_{i=0}^{n} a_i x^i \end{aligned}$$

we can *always* rewrite it as

$$\begin{aligned} p(x) &= a_n(x-z_n)(x-z_{n-1})(x-z_{n-2})\cdots(x-z_2)(x-z_1) \\ &\triangleq a_n \prod_{i=1}^{n}(x-z_i) \end{aligned}$$

where the points z_i are the polynomial roots, and they may be real or complex.

2.5 Complex Basics

This section introduces various notation and terms associated with complex numbers. As discussed above, complex numbers arise by introducing the square-root of -1 as a primitive new algebraic object among real numbers and manipulating it symbolically as if it were a real number itself:

$$\boxed{j \triangleq \sqrt{-1}}$$

Mathematicians and physicists often use i instead of j as $\sqrt{-1}$. The use of j is common in engineering where i is more often used for electrical current.

As mentioned above, for any negative number $c < 0$, we have

$$\sqrt{c} = \sqrt{(-1)(-c)} = j\sqrt{-c} = j\sqrt{|c|},$$

where $|c|$ denotes the absolute value of c. Thus, every square root of a negative number can be expressed as j times the square root of a positive number.

---

[4] Proofs for the fundamental theorem of algebra have a long history involving many of the great names in classical mathematics. The first known rigorous proof was by Gauss based on earlier efforts by Euler and Lagrange. (Gauss also introduced the term "complex number.") An alternate proof was given by Argand based on the ideas of d'Alembert. For a summary of the history, see `http://www-gap.dcs.st-and.ac.uk/~history/HistTopics/Fund_theorem_of_algebra.html` (the first Google search result for "fundamental theorem of algebra" in July of 2002).

By definition, we have

$$\begin{aligned} j^0 &= 1 \\ j^1 &= j \\ j^2 &= -1 \\ j^3 &= -j \\ j^4 &= 1 \\ &\cdots \end{aligned}$$

and so on. Thus, the sequence $x(n) \triangleq j^n$, $n = 0, 1, 2, \ldots$ is a periodic sequence with period 4, since $j^{n+4} = j^n j^4 = j^n$. (We'll learn later that the sequence j^n is a sampled complex sinusoid having frequency equal to one fourth the sampling rate.)

Every *complex number* z can be written as

$$\boxed{z = x + jy}$$

where x and y are real numbers. We call x the *real part* and y the *imaginary part*. We may also use the notation

$$\begin{aligned} \text{re}\{z\} &= x \qquad (\text{"the real part of } z = x + jy \text{ is } x\text{"}) \\ \text{im}\{z\} &= y \qquad (\text{"the imaginary part of } z = x + jy \text{ is } y\text{"}) \end{aligned}$$

Note that the real numbers are the subset of the complex numbers having a zero imaginary part ($y = 0$).

The rule for *complex multiplication* follows directly from the definition of the imaginary unit j:

$$\begin{aligned} z_1 z_2 &\triangleq (x_1 + jy_1)(x_2 + jy_2) \\ &= x_1 x_2 + jx_1 y_2 + jy_1 x_2 + j^2 y_1 y_2 \\ &= (x_1 x_2 - y_1 y_2) + j(x_1 y_2 + y_1 x_2) \end{aligned}$$

In some mathematics texts, complex numbers z are defined as ordered pairs of real numbers (x, y), and algebraic operations such as multiplication are defined more formally as operations on ordered pairs, *e.g.*, $(x_1, y_1) \cdot (x_2, y_2) \triangleq (x_1 x_2 - y_1 y_2, x_1 y_2 + y_1 x_2)$. However, such formality tends to obscure the underlying simplicity of complex numbers as a straightforward extension of real numbers to include $j \triangleq \sqrt{-1}$.

It is important to realize that complex numbers can be treated algebraically just like real numbers. That is, they can be added, subtracted, multiplied, divided, etc., using exactly the same rules of algebra (since both real and complex numbers are mathematical *fields*). It is often preferable to think of complex numbers as being the true and proper setting for algebraic operations, with real numbers being the limited subset for which $y = 0$.

2.6 The Complex Plane

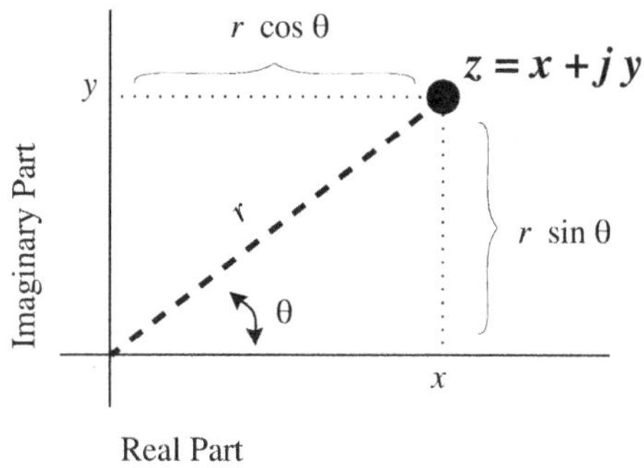

Figure 2.2: Plotting a complex number as a point in the complex plane.

We can plot any complex number $z = x + jy$ in a plane as an ordered pair (x, y), as shown in Fig. 2.2. A *complex plane* (or *Argand diagram*) is any 2D graph in which the horizontal axis is the *real part* and the vertical axis is the *imaginary part* of a complex number or function. As an example, the number j has coordinates $(0, 1)$ in the complex plane while the number 1 has coordinates $(1, 0)$.

Plotting $z = x + jy$ as the point (x, y) in the complex plane can be viewed as a plot in *Cartesian* or *rectilinear* coordinates. We can also express complex numbers in terms of *polar coordinates* as an ordered pair (r, θ), where r is the distance from the origin $(0, 0)$ to the number being plotted, and θ is the angle of the number relative to the positive real coordinate axis (the line defined by $y = 0$ and $x > 0$). (See Fig. 2.2.)

Using elementary geometry, it is quick to show that conversion from rectangular to polar coordinates is accomplished by the formulas

$$\begin{aligned} r &= \sqrt{x^2 + y^2} \\ \theta &= \tan^{-1}(y/x), \end{aligned}$$

where $\tan^{-1}(y/x)$ denotes the arctangent of y/x (the angle θ in radians whose tangent is $\tan(\theta) = y/x$). We will take θ in the range $-\pi$ to π (although we could choose any interval of length 2π radians, such as 0 to 2π, etc.).

The notation $\tan^{-1}(y, x)$ means "the angle whose tangent is y/x, taking the quadrant of the vector (x, y) into account." In Matlab and Octave, `atan2(y,x)` performs the "quadrant-sensitive" arctangent function. On the other hand, `atan(y/x)`, like the more traditional mathematical notation $\tan^{-1}(y/x)$ does not "know" the quadrant of (x, y), so it maps the entire real line to the interval $(-\pi/2, \pi/2)$. As a specific example, the angle of the vector $(x, y) = (1, 1)$ (in quadrant I) has the same tangent as the angle of $(x, y) = (-1, -1)$ (in quadrant III). Similarly, $(x, y) = (-1, 1)$ (quadrant II) yields the same tangent as $(x, y) = (1, -1)$ (quadrant IV).

The formula $r = \sqrt{x^2 + y^2}$ for converting rectangular coordinates to radius r, follows immediately from the *Pythagorean theorem*, while the $\theta = \tan^{-1}(y, x)$ follows from the definition of the tangent function itself.

Similarly, conversion from polar to rectangular coordinates is simply

$$\begin{aligned} x &= r\cos(\theta) \\ y &= r\sin(\theta). \end{aligned}$$

These follow immediately from the definitions of cosine and sine, respectively.

2.7 More Notation and Terminology

It's already been mentioned that the rectilinear coordinates of a complex number $z = x + jy$ in the complex plane are called the *real part* and *imaginary part*, respectively.

We also have special notation and various names for the *polar coordinates* (r, θ) of a complex number z:

$$\begin{aligned} r &\triangleq |z| = \sqrt{x^2 + y^2} \\ &= \textit{modulus, magnitude, absolute value, norm}, \text{ or } \textit{radial coordinate} \text{ of } z \\ \theta &\triangleq \angle z = \tan^{-1}(y/x) \\ &= \textit{angle, argument}, \text{ or } \textit{phase} \text{ of } z \end{aligned}$$

The *complex conjugate* of z is denoted $\overline{z}$ (or z^*) and is defined by

$$\boxed{\overline{z} \triangleq x - jy}$$

where, of course, $z \triangleq x + jy$.

In general, you can always obtain the complex conjugate of any expression by simply replacing j with $-j$. In the complex plane, this is a *vertical flip* about the real axis; *i.e.*, complex conjugation replaces each point in the complex plane by its *mirror image* on the other side of the x axis.

2.8 Elementary Relationships

From the above definitions, one can quickly verify

$$\begin{aligned} z + \overline{z} &= 2\,\mathrm{re}\{z\} \\ z - \overline{z} &= 2\,j\ \mathrm{im}\{z\} \\ z\overline{z} &= |z|^2. \end{aligned}$$

Let's verify the third relationship which states that a complex number multiplied by its conjugate is equal to its magnitude squared:

$$z\overline{z} \stackrel{\Delta}{=} (x + jy)(x - jy) = x^2 - (jy)^2 = x^2 + y^2 \stackrel{\Delta}{=} |z|^2 \tag{2.4}$$

2.9 Euler's Identity

Since $z = x + jy$ is the algebraic expression of z in terms of its rectangular coordinates, the corresponding expression in terms of its polar coordinates is

$$z = r\cos(\theta) + j\,r\sin(\theta).$$

There is another, more powerful representation of z in terms of its polar coordinates. In order to define it, we must introduce *Euler's identity*:

$$\boxed{e^{j\theta} = \cos(\theta) + j\sin(\theta)} \tag{2.5}$$

A proof of Euler's identity is given in the next chapter. Before, the only algebraic representation of a complex number we had was $z = x + jy$, which fundamentally uses Cartesian (rectilinear) coordinates in the complex plane. Euler's identity gives us an alternative representation in terms of polar coordinates in the complex plane:

$$\boxed{z = re^{j\theta}}$$

We'll call $re^{j\theta}$ the *polar form* of the complex number z, in contrast with the *rectangular form* $z = x + jy$. Polar form often simplifies algebraic manipulations of complex numbers, especially when they are multiplied together. Simple rules of exponents can often be used in place of messier trigonometric identities. In the case of two complex numbers being multiplied, we have

$$z_1 z_2 = \left(r_1 e^{j\theta_1}\right)\left(r_2 e^{j\theta_2}\right) = (r_1 r_2)\left(e^{j\theta_1} e^{j\theta_2}\right) = r_1 r_2 e^{j(\theta_1+\theta_2)}.$$

A corollary of Euler's identity is obtained by setting $\theta = \pi$ to get

$$e^{j\pi} + 1 = 0.$$

This has been called the "most beautiful formula in mathematics" due to the extremely simple form in which the fundamental constants $e, j, \pi, 1$, and 0, together with the elementary operations of addition, multiplication, exponentiation, and equality, all appear exactly once.

For another example of manipulating the polar form of a complex number, let's again verify $z\overline{z} = |z|^2$, as we did above in Eq. (2.4), but this time using polar form:

$$z\overline{z} = re^{j\theta}re^{-j\theta} = r^2e^0 = r^2 = |z|^2$$

As mentioned in §2.7, any complex expression can be conjugated by replacing j by $-j$ wherever it occurs. This implies $\overline{re^{j\theta}} = re^{-j\theta}$, as used above. The same result can be obtained by using Euler's identity to expand $re^{j\theta}$ into $r\cos(\theta)+jr\sin(\theta)$ and negating the imaginary part to obtain $\overline{re^{j\theta}} = r\cos(\theta) - jr\sin(\theta) = r\cos(-\theta) + jr\sin(-\theta) = re^{-j\theta}$, where we used also the fact that cosine is an *even* function ($\cos(-\theta) = \cos(\theta)$) while sine is *odd* ($\sin(-\theta) = -\sin(\theta)$).

We can now easily add a fourth line to that set of examples:

$$z/\overline{z} = \frac{re^{j\theta}}{re^{-j\theta}} = e^{j2\theta} = e^{j2\angle z}$$

Thus, $|z/\overline{z}| = 1$ for every $z \neq 0$.

Euler's identity can be used to derive formulas for sine and cosine in terms of $e^{j\theta}$:

$$\begin{aligned} e^{j\theta} + \overline{e^{j\theta}} &= e^{j\theta} + e^{-j\theta} \\ &= [\cos(\theta) + j\sin(\theta)] + [\cos(\theta) - j\sin(\theta)] \\ &= 2\cos(\theta) \end{aligned}$$

Similarly, $e^{j\theta} - \overline{e^{j\theta}} = 2j\sin(\theta)$, and we obtain the following classic identities:

$$\boxed{\cos(\theta) = \frac{e^{j\theta} + e^{-j\theta}}{2} \qquad \sin(\theta) = \frac{e^{j\theta} - e^{-j\theta}}{2j}}$$

2.10 De Moivre's Theorem

As a more complicated example of the value of the polar form, we'll prove *De Moivre's theorem:*

$$[\cos(\theta) + j\sin(\theta)]^n = \cos(n\theta) + j\sin(n\theta)$$

Working this out using sum-of-angle identities from trigonometry is laborious (see §3.13 on page 29 for details). However, using Euler's identity, De Moivre's theorem simply "falls out":

$$[\cos(\theta) + j\sin(\theta)]^n = \left[e^{j\theta}\right]^n = e^{j\theta n} = \cos(n\theta) + j\sin(n\theta)$$

Moreover, by the power of the method used to show the result, n can be any real number, not just an integer.

2.11 Conclusion

This chapter has covered just enough about complex numbers to enable us to talk about the discrete Fourier transform.

Manipulations of complex numbers in Matlab and Octave are illustrated in §I.1 on page 271.

To explore further the mathematics of complex variables, see any textbook such as Churchill [13] or LePage [35]. Topics not covered here, but which are important elsewhere in signal processing, include analytic functions, contour integration, analytic continuation, residue calculus, and conformal mapping.

2.12 Complex Number Problems

1. Find the roots of the following polynomials ($j = \sqrt{-1}$):

$(a)\ x^2 + 2x + 1$	$(d)\ 5x^2 - 2x + 1$	$(g)\ ax + b + j$
$(b)\ x^2 - 2x + 1$	$(e)\ x^2 + 9$	$(h)\ ax^2 + bx + j$
$(c)\ 5x^2 + 6x + 1$	$(f)\ x^3 + 2x^2 + x$	$(i)\ jx^2 + jx + j$

2. For the complex number $z = x + jy$, where x and y are real, find the
   (a) real part
   (b) imaginary part
   (c) modulus
   (d) phase
   (e) complex conjugate
   (f) reciprocal in rectangular form
   (g) reciprocal in polar form

3. Convert the following complex numbers to *Cartesian form* $x + jy$ (Example: $(1+j)(1-j) = 1 - j^2 = 2 = 2 + j0$):

(a) $(1+j)^2$ (e) $5e^{j\pi/2}$ (i) j^j

(b) $\frac{1+j}{1-j}$ (f) $\frac{\overline{(j^2)}}{(1+j)^2}$ (j) $\tan\left(\frac{1+j}{1-j}\right)$

(c) $\frac{a+jb}{c+jd}$ (g) $\ln(j)$ [hint: see (e)] (k) $e^{e^{j\theta}}$

(d) $e^{j\pi/6}$ (h) $(-1)^{1/4}$ (l) $\angle\left|\frac{1}{j+\frac{1}{j+\cdots}}\right|$

4. Convert the following expressions to *polar form* $re^{j\theta}$ (Example: $j = 0 + j1 = 1 \cdot e^{j\pi/2}$):

(a) $1+j$ (e) $x+jy$ (i) $\cos(\theta) + j\tan(\theta)$

(b) $1+j\sqrt{3}$ (f) $e^{j\pi}+1$ (j) $R\cos(\theta) + jR\sin(\theta)$

(c) $2+j2$ (g) $\sqrt{1+j}$ (k) $\sin(\theta) + j\cos(\theta)$

(d) $2\sqrt{3} - 2j$ (h) $(-1)^{1/10}$ (l) $\frac{1+j}{1-j}$

5. Consider the following derivation:

$$0 = \ln(1) = \ln(1^2) = \ln(-1)^2 = 2\ln(-1) \Longrightarrow \ln(-1) = 0$$

   (a) Is it true that $\ln(1) = \ln(-1) = 0$? If not, explain.

   (b) Find the set of all complex numbers which are equal to $\ln(-1)$.

6. Find all real numbers x satisfying the equation

$$x = j^j.$$

   Are there any solutions with a nonzero imaginary part?

7. Plot the complex numbers $e^{j2\pi k/8}$ in the complex plane for $k = 0, 1, \ldots, 7$. [Hint: use Euler's identity to find the Cartesian coordinates for each complex number.] On the same plot, draw the unit circle $|z| = 1$.

8. If a complex number z is multiplied by j, by how many degrees is z rotated in the complex plane? Is the rotation clockwise or counterclockwise? What is the rotation in radians?

9. Using DeMoivre's formula, find $(1+j)^{100}$ in polar form.

Chapter 3

Proof of Euler's Identity

This chapter outlines the proof of Euler's Identity, which is an important tool for working with complex numbers. It is one of the critical elements of the DFT definition that we need to understand.

3.1 Euler's Identity

Euler's identity (or "theorem" or "formula") is

$$e^{j\theta} = \cos(\theta) + j\sin(\theta) \qquad \text{(Euler's Identity)}$$

To "prove" this, we will first define what we mean by "$e^{j\theta}$". (The right-hand side, $\cos(\theta) + j\sin(\theta)$, is assumed to be understood.) Since e is just a particular real number, we only really have to explain what we mean by imaginary exponents. (We'll also see where e comes from in the process.) Imaginary exponents will be obtained as a generalization of real exponents. Therefore, our first task is to define exactly what we mean by a^x, where x is any real number, and $a > 0$ is any positive real number.

3.2 Positive Integer Exponents

The "original" definition of exponents which "actually makes sense" applies only to positive integer exponents:

$$\boxed{a^n \triangleq \underbrace{a\,a\,a\cdots a\,a}_{n \text{ times}}}$$

where $a > 0$ is real.

Generalizing this definition involves first noting its abstract mathematical properties, and then making sure these properties are preserved in the generalization.

3.3 Properties of Exponents

From the basic definition of positive integer exponents, we have

(1) $a^{n_1}a^{n_2} = a^{n_1+n_2}$

(2) $(a^{n_1})^{n_2} = a^{n_1 n_2}$

Note that property (1) implies property (2). We list them both explicitly for convenience below.

The Exponent Zero

How should we define a^0 in a manner consistent with the properties of integer exponents? Multiplying it by a gives

$$a^0 a = a^0 a^1 = a^{0+1} = a^1 = a$$

by property (1) of exponents. Solving $a^0 a = a$ for a^0 then gives

$$\boxed{a^0 = 1.}$$

Negative Exponents

What should a^{-1} be? Multiplying it by a gives, using property (1),

$$a^{-1} \cdot a = a^{-1}a^1 = a^{-1+1} = a^0 = 1.$$

Dividing through by a then gives

$$\boxed{a^{-1} = \frac{1}{a}.}$$

Similarly, we obtain

$$\boxed{a^{-M} = \frac{1}{a^M}}$$

for all integer values of M, *i.e.*, $\forall M \in \mathbf{Z}$.

3.4 Rational Exponents

A *rational* number is a real number that can be expressed as a ratio of two finite *integers*:

$$x = \frac{L}{M}, \quad L \in \mathbf{Z}, \quad M \in \mathbf{Z}$$

Applying property (2) of exponents, we have

$$a^x = a^{L/M} = \left(a^{\frac{1}{M}}\right)^L.$$

Thus, the only thing new is $a^{1/M}$. Since

$$\left(a^{\frac{1}{M}}\right)^M = a^{\frac{M}{M}} = a$$

we see that $a^{1/M}$ is the Mth root of a. This is sometimes written

$$\boxed{a^{\frac{1}{M}} \stackrel{\Delta}{=} \sqrt[M]{a}.}$$

The Mth root of a real (or complex) number is not unique. As we all know, square roots give two values (*e.g.*, $\sqrt{4} = \pm 2$). In the general case of Mth roots, there are M distinct values, in general. After proving Euler's identity, it will be easy to find them all (see §3.11 on page 27). As an example, $\sqrt[4]{1} = 1, -1, j$, and $-j$, since $1^4 = (-1)^4 = j^4 = (-j)^4 = 1$.

3.5 Real Exponents

The closest we can actually get to most real numbers is to compute a *rational number* that is as close as we need. It can be shown that rational numbers are *dense* in the real numbers; that is, between every two real numbers there is a rational number, and between every two rational numbers is a real number.[1] An *irrational* number can be defined as any real number having a non-repeating decimal expansion. For example, $\sqrt{2}$ is an irrational real number whose decimal expansion starts out as[2]

$$\sqrt{2} = 1.414213562373095048801688724209698078569671875376948073176679\ldots$$

[1]That the rationals are dense in the reals is easy to show using decimal expansions. Let r_1 and r_2 denote any two distinct, positive, irrational real numbers. Since r_1 and r_2 are distinct, their decimal expansions must differ in some digit, say the nth. Without loss of generality, assume $r_2 > r_1$. Form the rational number q_2 by zeroing all digits in the decimal expansion of r_2 after the nth. Then $r_2 > q_2 > r_1$, as needed. For two negative real numbers, we can negate them, use the same argument, and negate the result. For one positive and one negative real number, the rational number zero lies between them.

[2]This was computed via `N[Sqrt[2],60]` in `Mathematica`. Symbolic mathematics programs, such as `Mathematica`, `Maple` (offered as a `Matlab` extension), `maxima` (a free, GNU descendant of the original `Macsyma`, written in Common Lisp, and available at

Every truncated, rounded, or repeating expansion is a *rational* number. That is, it can be rewritten as an integer divided by another integer. For example,

$$1.414 = \frac{1414}{1000}$$

and, using $\overline{\text{overbar}}$ to denote the repeating part of a decimal expansion, a repeating example is as follows:

$$\begin{aligned} x &= 0.\overline{123} \\ \Rightarrow \quad 1000x &= 123.\overline{123} = 123 + x \\ \Rightarrow \quad 999x &= 123 \\ \Rightarrow \quad x &= \frac{123}{999} \end{aligned}$$

Other examples of irrational numbers include

$$\begin{aligned} \pi &= 3.1415926535897932384626433832795028841971693993751058209749\ldots \\ e &= 2.7182818284590452353602874713526624977572470936999595749669\ldots. \end{aligned}$$

Their decimal expansions do not repeat.

Let $\hat{x}_n$ denote the n-digit decimal expansion of an arbitrary real number x. Then $\hat{x}_n$ is a rational number (some integer over 10^n). We can say

$$\lim_{n\to\infty} \hat{x}_n = x.$$

That is, the *limit* of $\hat{x}_n$ as n goes to infinity is x.

Since $a^{\hat{x}_n}$ is defined for all n, we naturally define a^x as the following mathematical limit:

$$\boxed{a^x \triangleq \lim_{n\to\infty} a^{\hat{x}_n}}$$

We have now defined what we mean by *real* exponents.

---

http://maxima.sourceforge.net), GiNaC (the Maple-replacement used in the *Octave Symbolic Manipulation Toolbox*), or Yacas (another free, open-source program with similar goals as Mathematica), are handy tools for cranking out any number of digits in irrational numbers such as $\sqrt{2}$. In Yacas (as of Version 1.0.55), the syntax is

```
Precision(60)
N(Sqrt(2))
```

Of course, symbolic math programs can do much more than this, such as carrying out algebraic manipulations on polynomials and solving systems of symbolic equations in closed form.

3.6 A First Look at Taylor Series

Most "smooth" functions $f(x)$ can be expanded in the form of a *Taylor series expansion*:

$$f(x) = f(x_0) + \frac{f'(x_0)}{1}(x - x_0) + \frac{f''(x_0)}{1 \cdot 2}(x - x_0)^2 + \frac{f'''(x_0)}{1 \cdot 2 \cdot 3}(x - x_0)^3 + \cdots .$$

This can be written more compactly as

$$f(x) = \sum_{n=0}^{\infty} \frac{f^{(n)}(x_0)}{n!}(x - x_0)^n,$$

where '$n!$' is pronounced "n factorial". An informal derivation of this formula for $x_0 = 0$ is given in Appendix E. Clearly, since many derivatives are involved, a Taylor series expansion is only possible when the function is so smooth that it can be differentiated again and again. Fortunately for us, all audio signals are in that category, because hearing is bandlimited to below 20 kHz, and the audible spectrum of any sum of sinusoids is infinitely differentiable. (Recall that $\sin'(x) = \cos(x)$ and $\cos'(x) = -\sin(x)$, etc.). See §E.6 on page 237 for more about this point.

3.7 Imaginary Exponents

We may define *imaginary exponents* the same way that all sufficiently smooth real-valued functions of a real variable are generalized to the complex case—using Taylor series. A Taylor series expansion is just a polynomial (possibly of infinitely high order), and polynomials involve only addition, multiplication, and division. Since these elementary operations are also defined for complex numbers, any smooth function of a real variable $f(x)$ may be generalized to a function of a complex variable $f(z)$ by simply substituting the complex variable $z = x + jy$ for the real variable x in the Taylor series expansion of $f(x)$.

Let $f(x) \triangleq a^x$, where a is any positive real number and x is real. The Taylor series expansion about $x_0 = 0$ ("Maclaurin series"), generalized to the complex case is then

$$a^z \triangleq f(0) + f'(0)(z) + \frac{f''(0)}{2}z^2 + \frac{f'''(0)}{3!}z^3 + \cdots . \tag{3.1}$$

This is well defined, provided the series *converges* for every finite z (see Problem 8). We have $f(0) \triangleq a^0 = 1$, so the first term is no problem. But what is $f'(0)$? In other words, what is the derivative of a^x at $x = 0$? Once we find the successive derivatives of $f(x) \triangleq a^x$ at $x = 0$, we will have the definition of a^z for any complex z.

3.8 Derivatives of $f(x) = a^x$

Let's apply the definition of differentiation and see what happens:

$$\begin{aligned} f'(x_0) &\stackrel{\Delta}{=} \lim_{\delta\to 0}\frac{f(x_0+\delta)-f(x_0)}{\delta} \\ &\stackrel{\Delta}{=} \lim_{\delta\to 0}\frac{a^{x_0+\delta}-a^{x_0}}{\delta} = \lim_{\delta\to 0} a^{x_0}\frac{a^\delta-1}{\delta} = a^{x_0}\lim_{\delta\to 0}\frac{a^\delta-1}{\delta}. \end{aligned}$$

Since the limit of $(a^\delta - 1)/\delta$ as $\delta \to 0$ is less than 1 for $a = 2$ and greater than 1 for $a = 3$ (as one can show via direct calculations), and since $(a^\delta - 1)/\delta$ is a continuous function of a for $\delta > 0$, it follows that there exists a positive real number we'll call e such that for $a = e$ we get

$$\lim_{\delta\to 0}\frac{e^\delta-1}{\delta} \stackrel{\Delta}{=} 1.$$

For $a = e$, we thus have $(a^x)' = (e^x)' = e^x$.

So far we have proved that the derivative of e^x is e^x. What about a^x for other values of a? The trick is to write it as

$$a^x = e^{\ln(a^x)} = e^{x\ln(a)}$$

and use the chain rule,[3] where $\ln(a) \stackrel{\Delta}{=} \log_e(a)$ denotes the log-base-e of a.[4] Formally, the chain rule tells us how to differentiate a function of a function as follows:

$$\frac{d}{dx}f(g(x)) = f'(g(x))g'(x)$$

Evaluated at a particular point x_0, we obtain

$$\frac{d}{dx}f(g(x))|_{x=x_0} = f'(g(x_0))g'(x_0).$$

In this case, $g(x) = x\ln(a)$ so that $g'(x) = \ln(a)$, and $f(y) = e^y$ which is its own derivative. The end result is then $(a^x)' = \left(e^{x\ln a}\right)' = e^{x\ln(a)}\ln(a) = a^x\ln(a)$, *i.e.*,

$$\boxed{\frac{d}{dx}a^x = a^x\ln(a).}$$

[3] We will use the chain rule from calculus without proof. Note that the use of calculus is beyond the normal level of this book. Since calculus is only needed at this one point in the DFT-math story, the reader should not be discouraged if its usage seems like "magic". Calculus will not needed at all for practical applications of the DFT, such as spectrum analysis, discussed in Chapter 8.

[4] Logarithms are reviewed in Appendix F.

3.9 Back to e

Above, we defined e as the particular real number satisfying

$$\lim_{\delta\to 0}\frac{e^\delta - 1}{\delta} \triangleq 1$$

which gave us $(a^x)' = a^x$ when $a = e$. From this expression, we have, as $\delta \to 0$,

$$\begin{array}{rccl} & e^\delta - 1 & \to & \delta \\ \Rightarrow & e^\delta & \to & 1+\delta \\ \Rightarrow & e & \to & (1+\delta)^{1/\delta}, \end{array}$$

or

$$\boxed{e \triangleq \lim_{\delta\to 0}(1+\delta)^{1/\delta}.}$$

This is one way to define e. Another way to arrive at the same definition is to ask what logarithmic base e gives that the derivative of $\log_e(x)$ is $1/x$. We denote $\log_e(x)$ by $\ln(x)$.

Numerically, e is a transcendental number (a type of irrational number[5]), so its decimal expansion never repeats. The initial decimal expansion of e is given by[6]

$$e = 2.718281828459045235360287471352662497757247093 7\ldots.$$

Any number of digits can be computed from the formula $(1+\delta)^{1/\delta}$ by making δ sufficiently small.

3.10 $e^{j\theta}$

We've now defined a^z for any positive real number a and any complex number z. Setting $a = e$ and $z = j\theta$ gives us the special case we need for

[5]A number is said to be *transcendental* if it is not a root of any polynomial with integer coefficients, *i.e.*, it is not an *algebraic number* of any degree. (Rational numbers are algebraic numbers of degree 1; irrational numbers include transcendental numbers and algebraic numbers of degree greater than 1, such as $\sqrt{2}$ which is of degree 2.) See `http://mathworld.wolfram.com/TranscendentalNumber.html` for further discussion.

[6]In `Mathematica`, the first 50 digits of e may be computed by the expression `N[E,50]` ("evaluate numerically the reserved-constant `E` to 50 decimal places"). In the *Octave Symbolic Manipulation Toolbox* (part of Octave Forge), one may type "`digits(50); Exp(1)`".

Euler's identity. Since e^x is its own derivative, the Taylor series expansion for $f(x) = e^x$ is one of the simplest imaginable infinite series:

$$e^x = \sum_{n=0}^{\infty} \frac{x^n}{n!} = 1 + x + \frac{x^2}{2} + \frac{x^3}{3!} + \cdots$$

The simplicity comes about because $f^{(n)}(0) = 1$ for all n and because we chose to expand about the point $x = 0$. We of course define

$$e^{j\theta} \triangleq \sum_{n=0}^{\infty} \frac{(j\theta)^n}{n!} = 1 + j\theta - \frac{\theta^2}{2} - j\frac{\theta^3}{3!} + \cdots .$$

Note that all even order terms are real while all odd order terms are imaginary. Separating out the real and imaginary parts gives

$$\begin{aligned} \text{re}\left\{e^{j\theta}\right\} &= 1 - \theta^2/2 + \theta^4/4! - \cdots \\ \text{im}\left\{e^{j\theta}\right\} &= \theta - \theta^3/3! + \theta^5/5! - \cdots . \end{aligned}$$

Comparing the Maclaurin expansion for $e^{j\theta}$ with that of $\cos(\theta)$ and $\sin(\theta)$ proves Euler's identity. Recall from introductory calculus that

$$\begin{aligned} \frac{d}{d\theta}\cos(\theta) &= -\sin(\theta) \\ \frac{d}{d\theta}\sin(\theta) &= \cos(\theta) \end{aligned}$$

so that

$$\left.\frac{d^n}{d\theta^n}\cos(\theta)\right|_{\theta=0} = \begin{cases} (-1)^{n/2}, & n \text{ even} \\ 0, & n \text{ odd} \end{cases}$$

$$\left.\frac{d^n}{d\theta^n}\sin(\theta)\right|_{\theta=0} = \begin{cases} (-1)^{(n-1)/2}, & n \text{ odd} \\ 0, & n \text{ even.} \end{cases}$$

Plugging into the general Maclaurin series gives

$$\begin{aligned} \cos(\theta) &= \sum_{n=0}^{\infty} \frac{f^{(n)}(0)}{n!}\theta^n \\ &= \sum_{\substack{n\geq 0 \\ n \text{ even}}}^{\infty} \frac{(-1)^{n/2}}{n!}\theta^n \\ \sin(\theta) &= \sum_{\substack{n\geq 0 \\ n \text{ odd}}}^{\infty} \frac{(-1)^{(n-1)/2}}{n!}\theta^n . \end{aligned}$$

Separating the Maclaurin expansion for $e^{j\theta}$ into its even and odd terms (real and imaginary parts) gives

$$\begin{aligned} e^{j\theta} \triangleq \sum_{n=0}^{\infty} \frac{(j\theta)^n}{n!} &= \sum_{\substack{n\geq 0 \\ n \text{ even}}}^{\infty} \frac{(-1)^{n/2}}{n!}\theta^n + j \sum_{\substack{n\geq 0 \\ n \text{ odd}}}^{\infty} \frac{(-1)^{(n-1)/2}}{n!}\theta^n \\ &= \cos(\theta) + j\sin(\theta) \end{aligned}$$

thus proving Euler's identity.

3.11 Back to Mth Roots

As mentioned in §3.4 on page 21, there are M different numbers r which satisfy $r^M = a$ when M is a positive integer. That is, the Mth root of a, which is written as $a^{1/M}$, is not unique—there are M of them. How do we find them all? The answer is to consider *complex* numbers in *polar form*. By Euler's Identity, which we just proved, any number, real or complex, can be written in polar form as

$$z = re^{j\theta}$$

where $r \geq 0$ and $\theta \in [-\pi, \pi)$ are real numbers. Since, by Euler's identity, $e^{j2\pi k} = \cos(2\pi k) + j\sin(2\pi k) = 1$ for every integer k, we also have

$$z = re^{j\theta}e^{j2\pi k}.$$

Taking the Mth root gives

$$z^{\frac{1}{M}} = \left(re^{j\theta}e^{j2\pi k}\right)^{\frac{1}{M}} = r^{\frac{1}{M}}e^{j\frac{\theta}{M}}e^{j2\pi\frac{k}{M}} = r^{\frac{1}{M}}e^{j\frac{\theta+2\pi k}{M}}, \quad k \in \mathbf{Z}.$$

There are M different results obtainable using different values of k, *e.g.*, $k = 0, 1, 2, \ldots, M-1$. When $k = M$, we get the same thing as when $k = 0$. When $k = M+1$, we get the same thing as when $k = 1$, and so on, so there are only M distinct cases. Thus, we may define the kth Mth-root of $z = re^{j\theta}$ as

$$r^{\frac{1}{M}}e^{j\frac{\theta+2\pi k}{M}}, \quad k = 0, 1, 2, \ldots, M-1.$$

These are the M Mth-roots of the complex number $z = re^{j\theta}$.

3.12 Roots of Unity

Since $e^{j2\pi k} = 1$ for every integer k, we can write

$$1^{k/M} = e^{j2\pi k/M}, \quad k = 0, 1, 2, 3, \ldots, M-1.$$

These are the *Mth roots of unity*. The special case $k = 1$ is called a *primitive Mth root of unity*, since integer powers of it give all of the others:

$$e^{j2\pi k/M} = \left(e^{j2\pi/M}\right)^k$$

The Mth roots of unity are so frequently used that they are often given a special notation in the signal processing literature:

$$W_M^k \triangleq e^{j2\pi k/M}, \qquad k = 0, 1, 2, \ldots, M-1,$$

where W_M denotes a primitive Mth root of unity.[7] We may also call W_M a *generator* of the mathematical *group* consisting of the Mth roots of unity and their products.

We will learn later that the Nth roots of unity are used to generate all the sinusoids used by the length-N DFT and its inverse. The kth complex sinusoid used in a DFT of length N is given by

$$W_N^{kn} = e^{j2\pi kn/N} \triangleq e^{j\omega_k t_n} = \cos(\omega_k t_n) + j\sin(\omega_k t_n), \quad n = 0, 1, 2, \ldots, N-1,$$

where $\omega_k \triangleq 2\pi k/NT$, $t_n \triangleq nT$, and T is the sampling interval in seconds.

[7] Sometimes we see $W_M \triangleq e^{-j2\pi k/M}$, which is the complex conjugate of the definition we have used here. It is similarly a primitive Mth root, since powers of it will generate all other Mth roots.

3.13 Direct Proof of De Moivre's Theorem

In §2.10 on page 16, De Moivre's theorem was introduced as a consequence of Euler's identity:

$$\boxed{[\cos(\theta) + j\sin(\theta)]^n = \cos(n\theta) + j\sin(n\theta), \qquad \text{for all } n \in \mathbf{R}}$$

To provide some further insight into the "mechanics" of Euler's identity, we'll provide here a direct proof of De Moivre's theorem for *integer* n using mathematical induction and elementary trigonometric identities.

Proof: To establish the "basis" of our mathematical induction proof, we may simply observe that De Moivre's theorem is trivially true for $n = 1$. Now assume that De Moivre's theorem is true for some positive integer n. Then we must show that this implies it is also true for $n + 1$, *i.e.*,

$$[\cos(\theta) + j\sin(\theta)]^{n+1} = \cos[(n+1)\theta] + j\sin[(n+1)\theta]. \tag{3.2}$$

Since it is true by hypothesis that

$$[\cos(\theta) + j\sin(\theta)]^n = \cos(n\theta) + j\sin(n\theta),$$

multiplying both sides by $[\cos(\theta) + j\sin(\theta)]$ yields

$$\begin{aligned}
[\cos(\theta) + j\sin(\theta)]^{n+1} &= [\cos(n\theta) + j\sin(n\theta)] \cdot [\cos(\theta) + j\sin(\theta)] \\
&= [\cos(n\theta)\cos(\theta) - \sin(n\theta)\sin(\theta)] \\
&\quad + j\,[\sin(n\theta)\cos(\theta) + \cos(n\theta)\sin(\theta)]. \qquad (3.3)
\end{aligned}$$

From trigonometry, we have the following sum-of-angle identities:

$$\begin{aligned}
\sin(\alpha + \beta) &= \sin(\alpha)\cos(\beta) + \cos(\alpha)\sin(\beta) \\
\cos(\alpha + \beta) &= \cos(\alpha)\cos(\beta) - \sin(\alpha)\sin(\beta)
\end{aligned}$$

These identities can be proved using only arguments from classical geometry.[8] Applying these to the right-hand side of Eq. (3.3), with $\alpha = n\theta$ and $\beta = \theta$, gives Eq. (3.2), and so the induction step is proved. □

De Moivre's theorem establishes that integer powers of $[\cos(\theta)+j\sin(\theta)]$ lie on a circle of radius 1 (since $\cos^2(\phi) + \sin^2(\phi) = 1$, for all $\phi \in [-\pi, \pi]$). It therefore can be used to determine all N of the Nth roots of unity (see §3.12 above). However, no definition of e emerges readily from De Moivre's theorem, nor does it establish a definition for imaginary exponents (which we defined using Taylor series expansion in §3.7 above).

[8] See, for example, `http://www-spof.gsfc.nasa.gov/stargaze/Strig5.htm`.

3.14 Euler Identity Problems

1. What is the definition of e and where does it come from? Why is it a fundamental constant?

2. Derive the identities

$$\begin{aligned}\cos(a+b) &= \cos(a)\cos(b) - \sin(a)\sin(b)\\ \sin(a+b) &= \sin(a)\cos(b) + \cos(a)\sin(b)\end{aligned}$$

   using Euler's identity and the basic rule of exponents

$$e^{j(a+b)} = e^{ja}e^{jb}.$$

3. Derive the sum-of-angle identities of the previous problem using the methods of classical geometry.

4. Using the following Taylor series expansions, prove Euler's identity:

$$\begin{aligned}e^x &= 1 + x + \frac{x^2}{2!} + \frac{x^3}{3!} + \dots\\ \cos(\theta) &= 1 - \frac{\theta^2}{2!} + \frac{\theta^4}{4!} + \dots\\ \sin(\theta) &= \theta - \frac{\theta^3}{3!} + \frac{\theta^5}{5!} + \dots\end{aligned}$$

5. Use Euler's identity to prove *DeMoivre's theorem*:

$$[\cos(\theta) + j\sin(\theta)]^n = \cos(n\theta) + j\sin(n\theta)$$

6. Find the four 4th-roots of j. That is, find all four distinct values of $j^{\frac{1}{4}}$. Express the answers in polar form with angles between $-\pi$ and π.

7. If $z = x + jy$, where x and y are real, show that $|a^z| = |a^x|$ for all real $a > 0$.

8. Show that the Taylor expansion of a^z given in Eq. (3.1) on page 23 converges for every finite $z \in \mathbf{C}$ and for every finite real $a > 0$.

Chapter 4

Sinusoids and Exponentials

This chapter provides an introduction to sinusoids, exponentials, complex sinusoids, and various associated terminology, such as exponential decay-time "t_{60}", in-phase and quadrature sinusoidal components, analytic signals, positive and negative frequencies, and constructive and destructive interference. The fundamental importance of sinusoids in the analysis of linear time-invariant systems is introduced. We also look at circular motion expressed as the vector sum of in-phase and quadrature sinusoidal motions. Both continuous and discrete-time sinusoids are considered. In particular, a sampled complex sinusoid is generated by successive powers of any complex number z.

4.1 Sinusoids

A *sinusoid* is any function having the following form:

$$x(t) = A\sin(\omega t + \phi)$$

where t is the independent (real) variable, and the fixed parameters A, ω, and ϕ are all real constants. In audio applications we typically have

$$\begin{aligned}
A &= \text{peak amplitude (nonnegative)} \\
\omega &= \text{radian frequency (rad/sec)} \\
&= 2\pi f \ (f \text{ in Hz}) \\
t &= \text{time (sec)} \\
f &= \text{frequency (Hz)} \\
\phi &= \text{initial phase (radians)} \\
\omega t + \phi &= \text{instantaneous phase (radians).}
\end{aligned}$$

An example is plotted in Fig. 4.1 on page 33.

The term "peak amplitude" is often shortened to "amplitude," *e.g.*, "the amplitude of the tone was measured to be 5 Pascals." Strictly speaking, however, the amplitude of a signal x is its instantaneous value $x(t)$ at any time t. The peak amplitude A satisfies $|x(t)| \leq A$. The "instantaneous magnitude" or simply "magnitude" of a signal $x(t)$ is given by $|x(t)|$, and the peak magnitude is the same thing as the peak amplitude.

The "phase" of a sinusoid normally means the "initial phase", but in some contexts it might mean "instantaneous phase", so be careful. Another term for initial phase is *phase offset*.

Note that *Hz* is an abbreviation for *Hertz* which physically means *cycles per second*. You might also encounter the notation *cps* (or "c.p.s.") for cycles per second (still in use by physicists and formerly used by engineers as well).

Since the sine function is periodic with period 2π, the initial phase $\phi \pm 2\pi$ is indistinguishable from ϕ. As a result, we may restrict the range of ϕ to any length 2π interval. When needed, we will choose

$$-\pi \leq \phi < \pi,$$

i.e., $\phi \in [-\pi, \pi)$. You may also encounter the convention $\phi \in [0, 2\pi)$.

Note that the *radian frequency* ω is equal to the time derivative of the *instantaneous phase* of the sinusoid:

$$\frac{d}{dt}(\omega t + \phi) = \omega$$

This is also how the instantaneous frequency is defined when the phase is *time varying.* Let

$$\theta(t) \stackrel{\Delta}{=} \omega t + \phi(t)$$

denote the instantaneous phase of a sinusoid with a time-varying phase-offset $\phi(t)$. Then the instantaneous frequency is again given by the time derivative of the instantaneous phase:

$$\frac{d}{dt}[\omega t + \phi(t)] = \omega + \frac{d}{dt}\phi(t)$$

4.1.1 Example Sinusoids

Figure 4.1 plots the sinusoid $A\sin(2\pi ft + \phi)$, for $A = 10$, $f = 2.5$, $\phi = \pi/4$, and $t \in [0, 1]$. Study the plot to make sure you understand the effect of changing each parameter (amplitude, frequency, phase), and also note the definitions of "peak-to-peak amplitude" and "zero crossings."

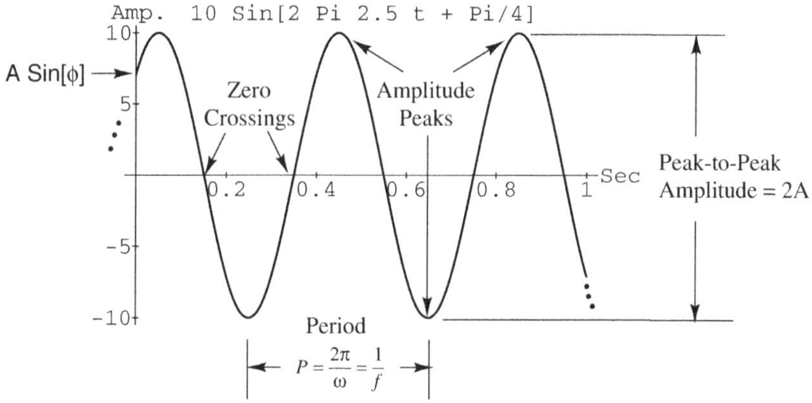

Figure 4.1: An example sinusoid.

A "tuning fork" vibrates approximately sinusoidally. An "A-440" tuning fork oscillates at 440 cycles per second. As a result, a tone recorded from an ideal A-440 tuning fork is a sinusoid at $f = 440$ Hz. The amplitude A determines how loud it is and depends on how hard we strike the tuning fork. The phase ϕ is set by exactly *when* we strike the tuning fork (and on our choice of when time 0 is). If we record an A-440 tuning fork on an analog tape recorder, the electrical signal recorded on tape is of the form

$$x(t) = A\sin(2\pi 440t + \phi).$$

As another example, the sinusoid at amplitude 1 and phase $\pi/2$ (90 degrees) is simply

$$x(t) = \sin(\omega t + \pi/2) = \cos(\omega t).$$

Thus, $\cos(\omega t)$ is a sinusoid at phase 90-degrees, while $\sin(\omega t)$ is a sinusoid at zero phase. Note, however, that we could just as well have defined $\cos(\omega t)$ to be the zero-phase sinusoid rather than $\sin(\omega t)$. It really doesn't matter, except to be consistent in any given usage. The concept of a "sinusoidal signal" is simply that it is equal to a sine or cosine function at some amplitude, frequency, and phase. It does not matter whether we choose sin() or cos() in the "official" definition of a sinusoid. You may encounter both definitions. Using sin() is nice since "sinusoid" naturally generalizes sin(). However, using cos() is nicer when defining a sinusoid to be the real part of a complex sinusoid (which we'll talk about in §4.3.11 on page 59).

4.1.2 Why Sinusoids are Important

Sinusoids arise naturally in a variety of ways:

One reason for the importance of sinusoids is that they are *fundamental in physics*. Many physical systems that *resonate* or *oscillate* produce quasi-sinusoidal motion. See *simple harmonic motion* in any freshman physics text for an introduction to this topic. The canonical example is the mass-spring oscillator.[1]

Another reason sinusoids are important is that they are *eigenfunctions of linear systems* (which we'll say more about in §4.1.4). This means that they are important in the analysis of *filters* such as reverberators, equalizers, certain (but not all) "audio effects", etc.

Perhaps most importantly, from the point of view of computer music research, is that the human *ear* is a kind of *spectrum analyzer*. That is, the cochlea of the inner ear physically splits sound into its (quasi) sinusoidal components. This is accomplished by the *basilar membrane* in the inner ear: a sound wave injected at the *oval window* (which is connected via the bones of the middle ear to the *ear drum*), travels along the basilar membrane inside the coiled cochlea. The membrane starts out thick and stiff, and gradually becomes thinner and more compliant toward its apex (the *helicotrema*). A stiff membrane has a high resonance frequency while a thin, compliant membrane has a low resonance frequency (assuming comparable mass per unit length, or at least less of a difference in mass than in compliance). Thus, as the sound wave travels, each frequency in the sound

[1] A mass-spring oscillator analysis is given at `http://ccrma.stanford.edu/~jos/filters/Mass_Spring_Oscillator_Analysis.html` (from the next book [65] in the music signal processing series).

resonates at a particular *place* along the basilar membrane. The highest audible frequencies resonate right at the entrance, while the lowest frequencies travel the farthest and resonate near the helicotrema. The membrane resonance effectively "shorts out" the signal energy at the resonant frequency, and it travels no further. Along the basilar membrane there are *hair cells* which "feel" the resonant vibration and transmit an increased firing rate along the auditory nerve to the brain. Thus, the ear is very literally a Fourier analyzer for sound, albeit nonlinear and using "analysis" parameters that are difficult to match exactly. Nevertheless, by looking at spectra (which display the amount of each sinusoidal frequency present in a sound), we are looking at a representation much more like what the brain receives when we hear.

4.1.3 In-Phase & Quadrature Sinusoidal Components

From the trig identity $\sin(A+B) = \sin(A)\cos(B) + \cos(A)\sin(B)$, we have

$$\begin{aligned} x(t) &\triangleq A\sin(\omega t + \phi) = A\sin(\phi + \omega t) \\ &= [A\sin(\phi)]\cos(\omega t) + [A\cos(\phi)]\sin(\omega t) \\ &\triangleq A_1\cos(\omega t) + A_2\sin(\omega t). \end{aligned}$$

From this we may conclude that every sinusoid can be expressed as the sum of a sine function (phase zero) and a cosine function (phase $\pi/2$). If the sine part is called the "in-phase" component, the cosine part can be called the "phase-quadrature" component. In general, "phase quadrature" means "90 degrees out of phase," *i.e.*, a relative phase shift of $\pm\pi/2$.

It is also the case that every sum of an in-phase and quadrature component can be expressed as a single sinusoid at some amplitude and phase. The proof is obtained by working the previous derivation backwards.

Figure 4.2 illustrates in-phase and quadrature components overlaid. Note that they only differ by a relative 90 degree phase shift.

4.1.4 Sinusoids at the Same Frequency

An important property of sinusoids at a particular frequency is that they are *closed* with respect to addition. In other words, if you take a sinusoid, make many copies of it, scale them all by different gains, delay them all by different time intervals, and add them up, you always get a sinusoid at the same original frequency. This is a nontrivial property. It obviously holds for any constant signal $x(t) = c$ (which we may regard as a sinusoid at frequency $f = 0$), but it is not obvious for $f \neq 0$ (see Fig. 4.2 and think about the sum of the two waveforms shown being precisely a sinusoid).

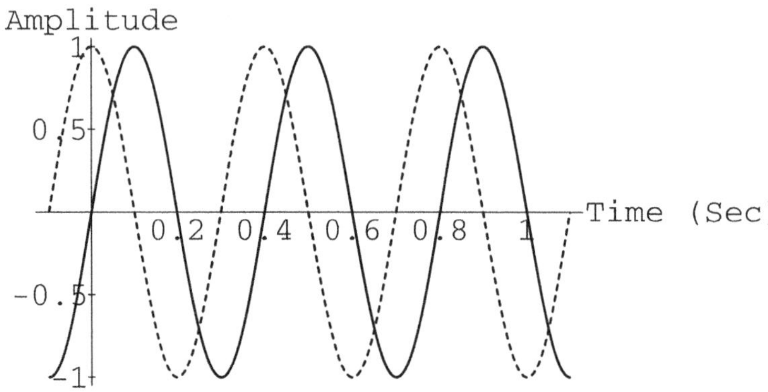

Figure 4.2: In-phase and quadrature sinusoidal components.

Since every linear, time-invariant (LTI[2]) system (filter) operates by copying, scaling, delaying, and summing its input signal(s) to create its output signal(s), it follows that when a sinusoid at a particular frequency is input to an LTI system, a sinusoid at that same frequency always appears at the output. Only the amplitude and phase can be changed by the system. We say that sinusoids are *eigenfunctions* of LTI systems. Conversely, if the system is nonlinear or time-varying, new frequencies are created at the system output.

To prove this important invariance property of sinusoids, we may simply express all scaled and delayed sinusoids in the "mix" in terms of their in-phase and quadrature components and then add them up. Here are the details in the case of adding two sinusoids having the same frequency. Let $x(t)$ be a general sinusoid at frequency ω:

$$x(t) \stackrel{\Delta}{=} A\sin(\omega t + \phi)$$

Now form $y(t)$ as the sum of two copies of $x(t)$ with arbitrary amplitudes and phase offsets:

$$\begin{aligned} y(t) &\stackrel{\Delta}{=} g_1 x(t-t_1) + g_2 x(t-t_2) \\ &= g_1 A\sin[\omega(t-t_1)+\phi] + g_2 A\sin[\omega(t-t_2)+\phi] \end{aligned}$$

[2]A system S is said to be *linear* if for any two input signals $x_1(t)$ and $x_2(t)$, we have $S[x_1(t)+x_2(t)] = S[x_1(t)] + S[x_2(t)]$. A system is said to be *time invariant* if $y(t) = S[x(t)]$ implies $S[x(t-\tau)] = y(t-\tau)$. This subject is developed in detail in the second book [65] of the music signal processing series, available on-line at `http://ccrma.stanford.edu/~jos/filters/`.

Focusing on the first term, we have

$$\begin{aligned}
g_1 A \sin[\omega(t - t_1) + \phi] &= g_1 A \sin[\omega t + (\phi - \omega t_1)] \\
&= [g_1 A \sin(\phi - \omega t_1)] \cos(\omega t) + \\
&\quad\; [g_1 A \cos(\phi - \omega t_1)] \sin(\omega t) \\
&\triangleq A_1 \cos(\omega t) + B_1 \sin(\omega t).
\end{aligned}$$

We similarly compute

$$g_2 A \sin[\omega(t - t_2) + \phi] = A_2 \cos(\omega t) + B_2 \sin(\omega t)$$

and add to obtain

$$y(t) = (A_1 + A_2) \cos(\omega t) + (B_1 + B_2) \sin(\omega t).$$

This result, consisting of one in-phase and one quadrature signal component, can now be converted to a single sinusoid at some amplitude and phase (and frequency ω), as discussed above.

4.1.5 Constructive and Destructive Interference

Sinusoidal signals are analogous to monochromatic laser light. You might have seen "speckle" associated with laser light, caused by destructive interference of multiple reflections of the light beam. In a room, the same thing happens with sinusoidal sound. For example, play a simple sinusoidal tone (*e.g.*, "A-440"—a sinusoid at frequency $f = 440$ Hz) and walk around the room with one ear plugged. If the room is reverberant you should be able to find places where the sound goes completely away due to destructive interference. In between such places (which we call "nodes" in the soundfield), there are "antinodes" at which the sound is louder by 6 dB (amplitude doubled—decibels (dB) are reviewed in Appendix F) due to constructive interference. In a *diffuse* reverberant soundfield,[3] the distance between nodes is on the order of a wavelength (the "correlation distance" within the random soundfield).

The way reverberation produces nodes and antinodes for sinusoids in a room is illustrated by the simple *comb filter*, depicted in Fig. 4.3.[4]

---

[3] For a definition, see
`http://ccrma.stanford.edu/~jos/pasp/Mean_Free_Path.html`

[4] Technically, Fig. 4.3 shows the *feedforward comb filter*, also called the "inverse comb filter" [72]. The longer names are meant to distinguish it from the *feedback comb filter*, in which the delay output is fed *back* around the delay line and summed with the delay input instead of the input being fed *forward* around the delay line and summed with its output. (A diagram and further discussion, in-

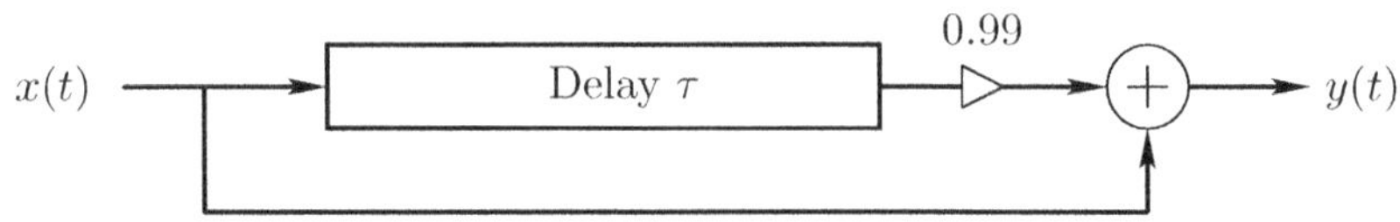

Figure 4.3: A *comb filter* with input signal $x(t)$, output signal $y(t)$, and delayed-input coefficient 0.99. The output signal is given by $y(t) = x(t) + 0.99\,x(t-\tau)$.

Since the comb filter is linear and time-invariant, its response to a sinusoid must be sinusoidal (see previous section). The feedforward path has gain 1, and the delayed signal is scaled by 0.99. With the delay set to one period, the sinusoid coming out of the delay line *constructively interferes* with the sinusoid from the feed-forward path, and the output amplitude is therefore $1 + 0.99 = 1.99$. In the opposite extreme case, with the delay set to *half* a period, the unit-amplitude sinusoid coming out of the delay line *destructively interferes* with the sinusoid from the feed-forward path, and the output amplitude therefore drops to $|1 - 0.99| = 0.01$.

Consider a fixed delay of τ seconds for the delay line in Fig. 4.3. Constructive interference happens at all frequencies for which an *exact integer* number of periods fits in the delay line, *i.e.*, $f\tau = 0, 1, 2, 3, \ldots$, or $f = n/\tau$, for $n = 0, 1, 2, 3, \ldots$. On the other hand, destructive interference happens at all frequencies for which there is an *odd number of half-periods*, *i.e.*, the number of periods in the delay line is an integer plus a half: $f\tau = 1.5, 2.5, 3.5$, etc., or, $f = (n + 1/2)/\tau$, for $n = 0, 1, 2, 3, \ldots$. It is quick to verify that frequencies of constructive interference alternate with frequencies of destructive interference, and therefore the *amplitude response* of the comb filter (a plot of gain versus frequency) looks as shown in Fig. 4.4.

The amplitude response of a comb filter has a "comb" like shape, hence the name.[5] It looks even more like a comb on a dB amplitude scale, as shown in Fig. 4.5. A dB scale is more appropriate for audio applications, as discussed in Appendix F. Since the minimum gain is $1 - 0.99 = 0.01$, the nulls in the response reach down to -40 dB; since the maximum gain is $1 + 0.99 \approx 2$, the maximum in dB is about 6 dB. If the feedforward gain were increased from 0.99 to 1, the nulls would extend, in principle, to minus infinity, corresponding to a gain of zero (complete cancellation). Negating

cluding how time-varying comb filters create a *flanging effect*, can be found at `http://ccrma.stanford.edu/~jos/pasp/Feedback_Comb_Filters.html`.) The frequency response of the feedforward comb filter is the inverse of that of the feedback comb filter (one can cancel the effect of the other), hence the name "inverse comb filter." Frequency-response analysis of digital filters is developed in [65] (available online).

[5]While there is no reason it should be obvious at this point, the comb-filter gain varies in fact sinusoidally as a function of frequency.

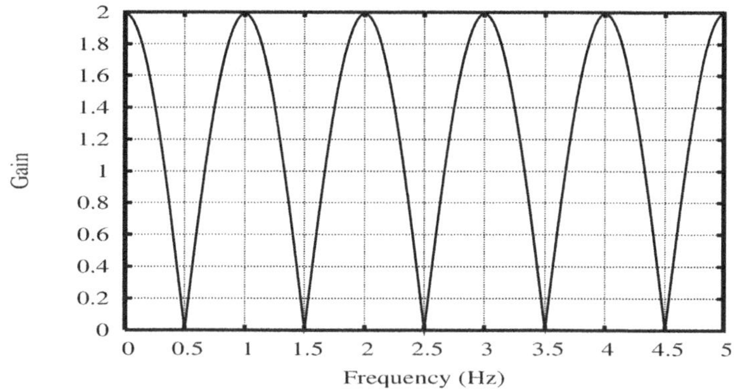

Figure 4.4: Comb filter amplitude response when delay $\tau = 1$ sec.

the feedforward path would shift the curve left (or right) by 1/2 Hz, placing a minimum at dc[6] instead of a peak.

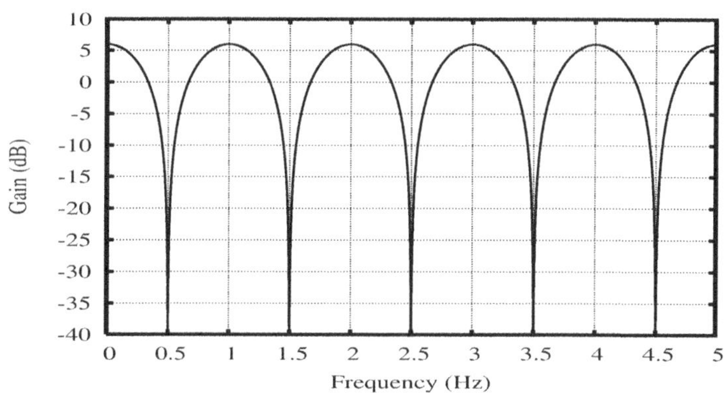

Figure 4.5: Comb filter amplitude response *in dB*.

4.1.6 Sinusoid Magnitude Spectra

A sinusoid's frequency content may be graphed in the *frequency domain* as shown in Fig. 4.6.

An example of a particular sinusoid graphed in Fig. 4.6 is given by

$$x(t) = \cos(\omega_x t) = \frac{1}{2}e^{j\omega_x t} + \frac{1}{2}e^{-j\omega_x t}$$

[6] "dc" means "direct current" and is an electrical engineering term for "frequency 0".

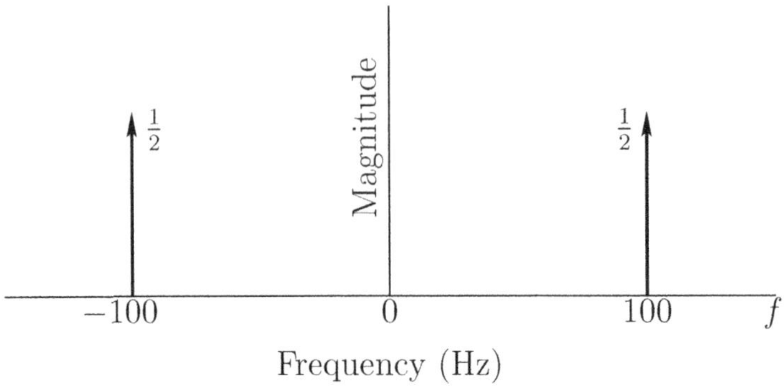

Figure 4.6: Spectral magnitude representation of a unit-amplitude sinusoid at frequency 100 Hz such as $\cos(200\pi t)$ or $\sin(200\pi t)$. (Phase is not shown.)

where

$$\omega_x = 2\pi 100.$$

That is, this sinusoid has amplitude 1, frequency 100 Hz, and phase zero (or $\pi/2$, if $\sin(\omega_x t)$ is defined as the zero-phase case).

Figure 4.6 can be viewed as a graph of the *magnitude spectrum* of $x(t)$, or its *spectral magnitude representation* [42]. Note that the spectrum consists of two components with amplitude 1/2, one at frequency 100 Hz and the other at frequency −100 Hz.

Phase is not shown in Fig. 4.6 at all. The phase of the components could be written simply as labels next to the magnitude arrows, or the magnitude arrows can be rotated "into or out of the page" by the appropriate phase angle, as illustrated in Fig. 4.16 on page 56.

4.2 Exponentials

The canonical form of an exponential function, as typically used in signal processing, is

$$a(t) = Ae^{-t/\tau}, \quad t \geq 0$$

where τ is called the *time constant* of the exponential. A is the peak amplitude, as before. The time constant is the time it takes to decay by $1/e$, *i.e.*,

$$\frac{a(\tau)}{a(0)} = \frac{1}{e}.$$

A normalized exponential decay is depicted in Fig. 4.7.

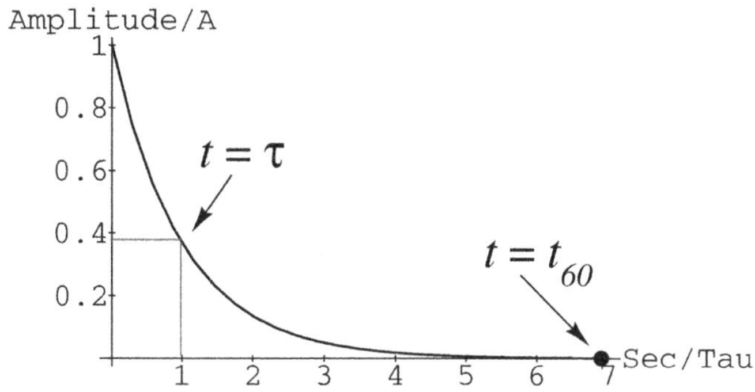

Figure 4.7: The decaying exponential $Ae^{-t/\tau}$, normalized to unit amplitude ($e^{-t/\tau}$).

4.2.1 Why Exponentials are Important

Exponential *decay* occurs naturally when a quantity is decaying at a rate which is proportional to how much is left. In nature, all *linear resonators*, such as musical instrument strings and woodwind bores, exhibit exponential decay in their response to a momentary excitation. As another example, reverberant energy in a room decays exponentially after the direct sound stops. Essentially all *undriven oscillations* decay exponentially (provided they are linear and time-invariant). Undriven means there is no ongoing source of driving energy. Examples of undriven oscillations include the vibrations of a tuning fork, struck or plucked strings, a marimba or xylophone bar, and so on. Examples of driven oscillations include horns, woodwinds, bowed strings, and voice. Driven oscillations must be periodic while undriven oscillations normally are not, except in idealized cases.

Exponential *growth* occurs when a quantity is increasing at a rate proportional to the current amount. Exponential growth is *unstable* since nothing can grow exponentially forever without running into some kind of limit. Note that a positive time constant corresponds to exponential decay, while a negative time constant corresponds to exponential growth. In signal processing, we almost always deal exclusively with exponential decay (positive time constants).

Exponential growth and decay are illustrated in Fig. 4.8.

4.2.2 Audio Decay Time (T60)

In audio, a decay by $1/e$ is not enough to become inaudible, unless the starting amplitude was extremely small. In *architectural acoustics* (which

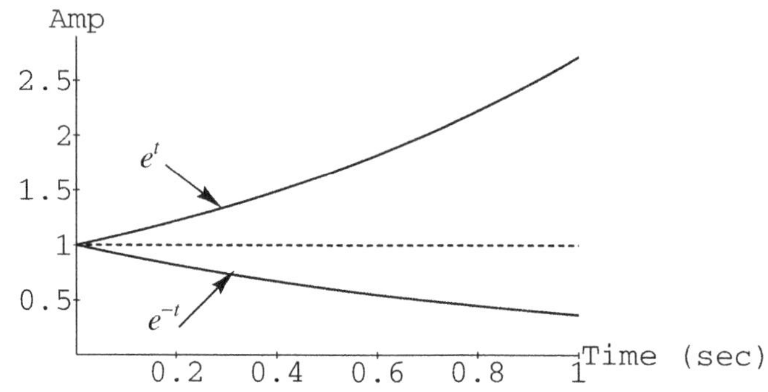

Figure 4.8: Growing and decaying exponentials.

includes the design of concert halls), a more commonly used measure of decay is "t_{60}" (or T60), which is defined as the *time to decay by* 60 *dB*.[7] That is, t_{60} is obtained by solving the equation

$$\frac{a(t_{60})}{a(0)} = 10^{-60/20} = 0.001.$$

Using the definition of the exponential $a(t) = Ae^{-t/\tau}$, we find

$$\boxed{t_{60} = \ln(1000)\tau \approx 6.91\tau}$$

Thus, t_{60} is about seven time constants. See where t_{60} is marked on Fig. 4.7 compared with τ.

4.3 Complex Sinusoids

Recall Euler's Identity,

$$e^{j\theta} = \cos(\theta) + j\sin(\theta).$$

Multiplying this equation by $A \geq 0$ and setting $\theta = \omega t + \phi$, where t is time in seconds, ω is radian frequency, and ϕ is a phase offset, we obtain what we call the *complex sinusoid*:

$$s(t) \stackrel{\Delta}{=} Ae^{j(\omega t+\phi)} = A\cos(\omega t + \phi) + jA\sin(\omega t + \phi)$$

[7]Recall that a gain factor g is converted to *decibels* (dB) by the formula $20\log_{10}(g)$. See §F.2 on page 241 for a review.

Thus, a complex sinusoid consists of an "in-phase" component for its real part, and a "phase-quadrature" component for its imaginary part. Since $\sin^2(\theta) + \cos^2(\theta) = 1$, we have

$$|s(t)| \stackrel{\Delta}{=} \sqrt{\mathrm{re}^2\{s(t)\} + \mathrm{im}^2\{s(t)\}} \equiv A.$$

That is, the complex sinusoid has a *constant modulus* (*i.e.*, a constant complex magnitude). (The symbol "$\equiv$" means "identically equal to," *i.e.*, for all t.) The instantaneous phase of the complex sinusoid is

$$\angle s(t) = \omega t + \phi.$$

The derivative of the instantaneous phase of the complex sinusoid gives its instantaneous frequency

$$\frac{d}{dt}\angle s(t) = \omega = 2\pi f.$$

4.3.1 Circular Motion

Since the modulus of the complex sinusoid is constant, it must lie on a *circle* in the complex plane. For example,

$$x(t) = e^{j\omega t}$$

traces out *counter-clockwise* circular motion along the unit circle in the complex plane as t increases, while

$$\overline{x(t)} = e^{-j\omega t}$$

gives *clockwise* circular motion.

We may call a complex sinusoid $e^{j\omega t}$ a *positive-frequency sinusoid* when $\omega > 0$. Similarly, we may define a complex sinusoid of the form $e^{-j\omega t}$, with $\omega > 0$, to be a *negative-frequency sinusoid.* Note that a positive- or negative-frequency sinusoid is necessarily complex.

4.3.2 Projection of Circular Motion

Interpreting the real and imaginary parts of the complex sinusoid,

$$\begin{aligned}\mathrm{re}\left\{e^{j\omega t}\right\} &= \cos(\omega t)\\ \mathrm{im}\left\{e^{j\omega t}\right\} &= \sin(\omega t),\end{aligned}$$

in the complex plane, we see that *sinusoidal motion is the projection of circular motion onto any straight line.* Thus, the sinusoidal motion $\cos(\omega t)$

is the projection of the circular motion $e^{j\omega t}$ onto the x (real-part) axis, while $\sin(\omega t)$ is the projection of $e^{j\omega t}$ onto the y (imaginary-part) axis.

Figure 4.9 shows a plot of a complex sinusoid versus time, along with its projections onto coordinate planes. This is a 3D plot showing the z-plane versus time. The axes are the real part, imaginary part, and time. (Or we could have used magnitude and phase versus time.)

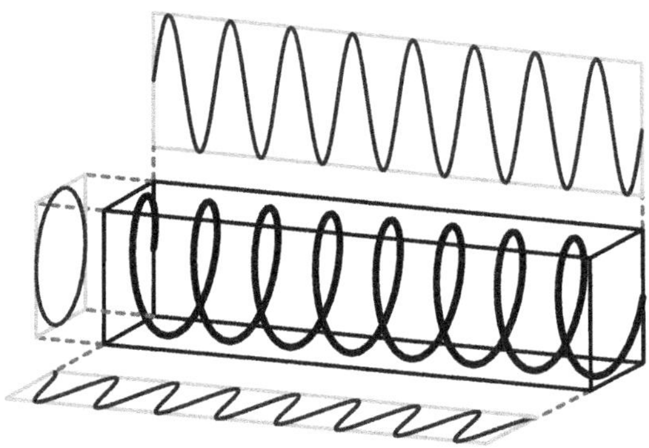

Figure 4.9: A complex sinusoid and its projections.

Note that the left projection (onto the z plane) is a circle, the lower projection (real-part vs. time) is a cosine, and the upper projection (imaginary-part vs. time) is a sine. A point traversing the plot projects to uniform circular motion in the z plane, and sinusoidal motion on the two other planes.

4.3.3 Positive and Negative Frequencies

In §2.9, we used Euler's Identity to show

$$\begin{aligned}\cos(\theta) &= \frac{e^{j\theta}+e^{-j\theta}}{2}\\ \sin(\theta) &= \frac{e^{j\theta}-e^{-j\theta}}{2j}\end{aligned}$$

Setting $\theta = \omega t + \phi$, we see that both sine and cosine (and hence all real sinusoids) consist of a sum of equal and opposite circular motion. Phrased differently, every real sinusoid consists of an equal contribution of positive and negative frequency components. This is true of all real signals. When we get to spectrum analysis, we will find that every real signal contains equal amounts of positive and negative frequencies, *i.e.*, if $X(\omega)$ denotes the spectrum of the real signal $x(t)$, we will always have $|X(-\omega)| = |X(\omega)|$.

Note that, mathematically, the complex sinusoid $Ae^{j(\omega t+\phi)}$ is really *simpler* and *more basic* than the real sinusoid $A\sin(\omega t + \phi)$ because $e^{j\omega t}$ consists of one frequency ω while $\sin(\omega t)$ really consists of two frequencies ω and $-\omega$. We may think of a real sinusoid as being the sum of a positive-frequency and a negative-frequency complex sinusoid, so in that sense real sinusoids are "twice as complicated" as complex sinusoids. Complex sinusoids are also nicer because they have a *constant modulus*. "Amplitude envelope detectors" for complex sinusoids are trivial: just compute the square root of the sum of the squares of the real and imaginary parts to obtain the *instantaneous peak amplitude* at any time. Frequency demodulators are similarly trivial: just differentiate the phase of the complex sinusoid to obtain its *instantaneous frequency*. It should therefore come as no surprise that signal processing engineers often prefer to convert real sinusoids into complex sinusoids (by filtering out the negative-frequency component) before processing them further.

4.3.4 Plotting Complex Sinusoids versus Frequency

As discussed in the previous section, we regard the signal

$$x(t) = A_x e^{j\omega_x t}$$

as a *positive-frequency sinusoid* when $\omega_x > 0$. In a manner analogous to spectral magnitude plots (discussed in §4.1.6 on page 39), we can plot this complex sinusoid over a frequency axis as a vertical line of length A_x at the point $\omega = \omega_x$, as shown in Fig. 4.10. Such a plot of amplitude versus frequency may be called a *spectral plot*, or *spectral representation* [42] of the (zero-phase) complex sinusoid.

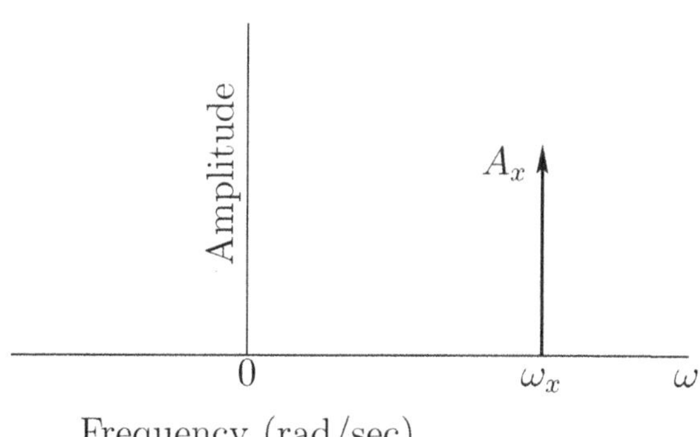

Figure 4.10: Spectral plot of a complex sinusoid $A_x e^{j\omega_x t}$.

More generally, however, a complex sinusoid has both an amplitude and a *phase* (or, equivalently, a *complex amplitude*):

$$x(t) = \left(A_x e^{j\theta_x}\right) e^{j\omega_x t}$$

To accommodate the phase angle θ_x in spectral plots, the plotted vector may be rotated by the angle θ_x in the plane orthogonal to the frequency axis passing through ω_x, as done in Fig. 4.16b below (p. 56) for phase angles $\theta_x = \pm\pi/2$.

4.3.5 Sinusoidal Amplitude Modulation (AM)

It is instructive to study the *modulation* of one sinusoid by another. In this section, we will look at sinusoidal *Amplitude Modulation (AM)*. The general AM formula is given by

$$x_\alpha(t) = [1 + \alpha \cdot a_m(t)] \cdot A_c \sin(\omega_c t + \phi_c),$$

where (A_c, ω_c, ϕ_c) are parameters of the sinusoidal *carrier wave*, $\alpha \in [0, 1]$ is called the *modulation index* (or *AM index*), and $a_m(t) \in [-1, 1]$ is the *amplitude modulation signal*. In AM radio broadcasts, $a_m(t)$ is the audio signal being transmitted (usually bandlimited to less than 10 kHz), and ω_c is the channel center frequency that one dials up on a radio receiver. The modulated signal $x_\alpha(t)$ can be written as the sum of the unmodulated carrier wave plus the product of the carrier wave and the modulating wave:

$$x_\alpha(t) = x_0(t) + \alpha \cdot a_m(t) \cdot A_c \sin(\omega_c t + \phi_c) \tag{4.1}$$

In the case of *sinusoidal* AM, we have

$$a_m(t) = \sin(\omega_m t + \phi_m). \tag{4.2}$$

Periodic amplitude modulation of this nature is often called the *tremolo effect* when $\omega_m < 20\pi$ or so (< 10 Hz).

Let's analyze the second term of Eq. (4.1) for the case of sinusoidal AM with $\alpha = 1$ and $\phi_m = \phi_c = 0$:

$$x_m(t) \triangleq \sin(\omega_m t)\sin(\omega_c t) \tag{4.3}$$

An example waveform is shown in Fig. 4.11 for $f_c = 100$ Hz and $f_m = 10$ Hz. Such a signal may be produced on an analog synthesizer by feeding two differently tuned sinusoids to a *ring modulator*, which is simply a "four-quadrant multiplier" for analog signals.

When ω_m is small (say less than 20π radians per second, or 10 Hz), the signal $x_m(t)$ is heard as a "beating sine wave" with $\omega_m/\pi = 2f_m$ beats per

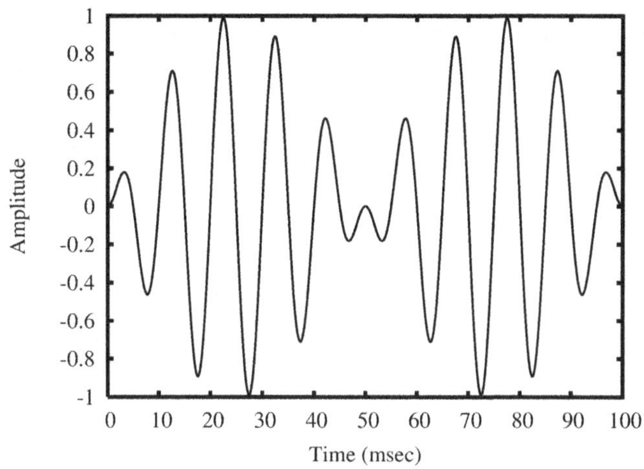

Figure 4.11: Sinusoidal amplitude modulation as in Eq. (4.3)—time waveform.

second. The beat rate is twice the modulation frequency because both the positive and negative peaks of the modulating sinusoid cause an "amplitude swell" in $x_m(t)$. (One period of modulation—$1/f_m$ seconds—is shown in Fig. 4.11.) The sign inversion during the negative peaks is not normally audible.

Recall the trigonometric identity for a sum of angles:

$$\cos(A+B) = \cos(A)\cos(B) - \sin(A)\sin(B)$$

Subtracting this from $\cos(A-B) = \cos(A)\cos(B) + \sin(A)\sin(B)$ leads to the identity

$$\sin(A)\sin(B) = \frac{\cos(A-B) - \cos(A+B)}{2}.$$

Setting $A = \omega_m t$ and $B = \omega_c t$ gives us an alternate form for our "ring-modulator output signal":

$$x_m(t) \stackrel{\Delta}{=} \sin(\omega_m t)\sin(\omega_c t) = \frac{\cos[(\omega_m - \omega_c)t] - \cos[(\omega_m + \omega_c)t]}{2} \tag{4.4}$$

These two sinusoidal components at the *sum and difference frequencies* of the modulator and carrier are called *side bands* of the carrier wave at frequency ω_c (since typically $\omega_c \gg \omega_m > 0$).

Equation (4.3) expresses $x_m(t)$ as a "beating sinusoid", while Eq. (4.4) expresses as it two *unmodulated* sinusoids at frequencies $\omega_c \pm \omega_m$. Which case do we hear?

It turns out we hear $x_m(t)$ as two separate tones (Eq. (4.4)) whenever the side bands are *resolved* by the ear. As mentioned in §4.1.2, the ear performs a "short time Fourier analysis" of incoming sound (the basilar membrane in the cochlea acts as a mechanical filter bank). The *resolution* of this filterbank—its ability to discern two separate spectral peaks for two sinusoids closely spaced in frequency—is determined by the *critical bandwidth of hearing* [43, 73, 84]. A critical bandwidth is roughly 15-20% of the band's center-frequency, over most of the audio range [68]. Thus, the side bands in sinusoidal AM are heard as separate tones when they are both in the audio range and separated by at least one critical bandwidth. When they are well inside the same critical band, "beating" is heard. In between these extremes, near separation by a critical-band, the sensation is often described as "roughness" [27].

Example AM Spectra

Equation (4.4) can be used to write down the spectral representation of $x_m(t)$ by inspection, as shown in Fig. 4.12. In the example of Fig. 4.12, we have $f_c = 100$ Hz and $f_m = 20$ Hz, where, as always, $\omega = 2\pi f$. For comparison, the spectral magnitude of an *unmodulated* 100 Hz sinusoid is shown in Fig. 4.6 on page 40. Note in Fig. 4.12 how each of the two sinusoidal components at ± 100 Hz have been "split" into two "side bands", one 20 Hz higher and the other 20 Hz lower, that is, $\pm 100 \pm 20 = \{-120, -80, 80, 120\}$. Note also how the *amplitude* of the split component is divided equally among its two side bands.

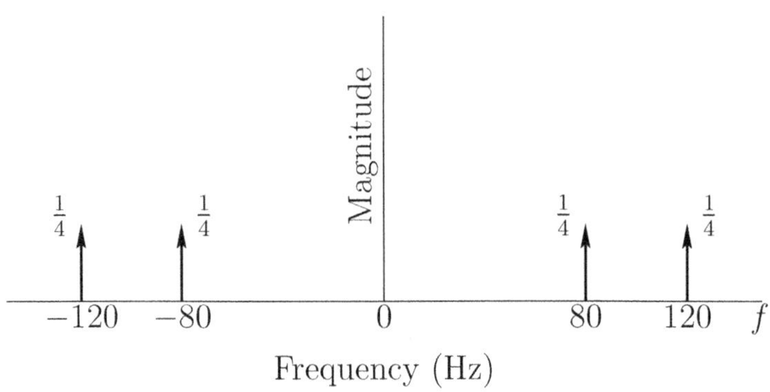

Figure 4.12: Spectral magnitude representation of the sinusoidally modulated sinusoid $\sin(40\pi t)\sin(200\pi t)$ defined in Eq. (4.3). Phase is not shown.

Recall that $x_m(t)$ was defined as the *second term* of Eq. (4.1). The first term is simply the original unmodulated signal. Therefore, we have effectively been considering AM with a "very large" modulation index. In the more general case of Eq. (4.1) with $a_m(t)$ given by Eq. (4.2), the magnitude of the spectral representation appears as shown in Fig. 4.13.

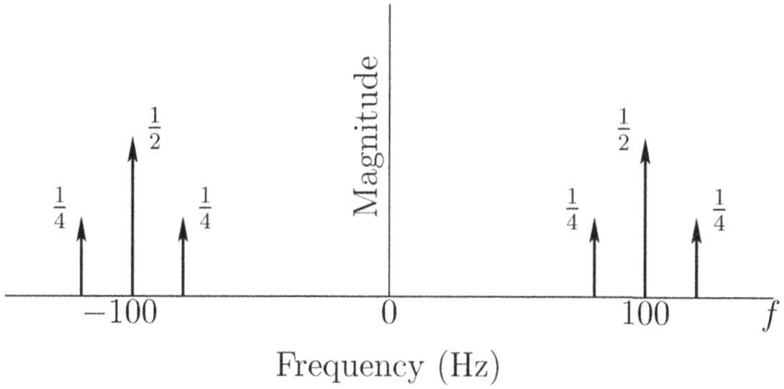

Figure 4.13: Spectral representation of the sinusoidally modulated sinusoid $[1+\sin(40\pi t)]\sin(200\pi t)$ from Eq. (4.1), with $\alpha = 1$, and $a_m(t)$ given by Eq. (4.2).

4.3.6 Sinusoidal Frequency Modulation (FM)

Frequency Modulation (FM) is well known as the broadcast signal format for FM radio. It is also the basis of the first commercially successful method for digital sound synthesis. Invented by John Chowning [12], it was the method used in the the highly successful Yamaha DX-7 synthesizer, and later the Yamaha OPL chip series, which was used in all "SoundBlaster compatible" multimedia sound cards for many years. At the time of this writing, descendants of the OPL chips remain the dominant synthesis technology for "ring tones" in cellular telephones.

A general formula for frequency modulation of one sinusoid by another can be written as

$$x(t) = A_c \cos[\omega_c t + \phi_c + A_m \sin(\omega_m t + \phi_m)], \tag{4.5}$$

where the parameters (A_c, ω_c, ϕ_c) describe the *carrier sinusoid*, while the parameters (A_m, ω_m, ϕ_m) specify the *modulator* sinusoid. Note that, strictly speaking, it is not the frequency of the carrier that is modulated sinusoidally, but rather the *instantaneous phase* of the carrier. Therefore, *phase modulation* would be a better term (which is in fact used). Potential confusion aside, any modulation of phase implies a modulation of frequency, and vice versa, since the instantaneous frequency is always defined as the time-derivative of the instantaneous phase. In this book, only phase modulation will be considered, and we will call it FM, following common practice.[8]

Figure 4.14 shows a unit generator patch diagram [40] for brass-like FM synthesis. For brass-like sounds, the modulation amount increases with the amplitude of the signal. In the patch, note that the amplitude envelope for the carrier oscillator is scaled and also used to control amplitude of the modulating oscillator.

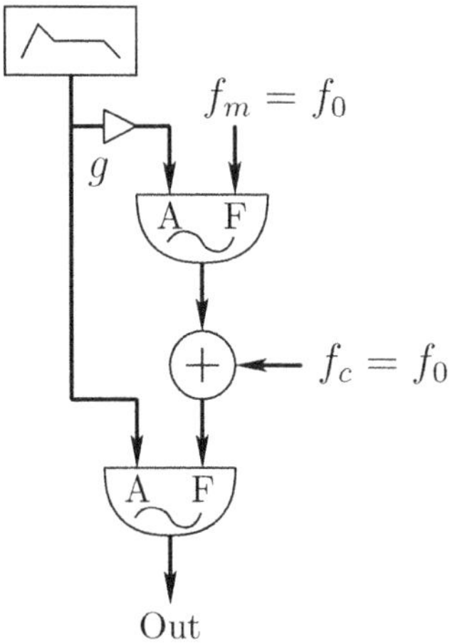

Figure 4.14: Unit generator diagram for simple FM brass synthesis.

It is well known that sinusoidal frequency-modulation of a sinusoid creates sinusoidal components that are uniformly spaced in frequency by multiples of the modulation frequency, with amplitudes given by the *Bessel functions* of the first kind [12]. As a special case, frequency-modulation of a sinusoid by itself generates a harmonic spectrum in which the kth harmonic amplitude is proportional to $J_k(\beta)$, where k is the *order* of the Bessel function and β is the *FM index*. We will derive this in the next section.[9]

[8] An important variant of FM called *feedback FM*, in which a single oscillator phase-modulates itself, simply does not work if true frequency modulation is implemented.

[9] The mathematical derivation of FM spectra is included here as a side note. No further use will be made of it in this book.

Bessel Functions

The *Bessel functions of the first kind* may be defined as the coefficients $J_k(\beta)$ in the two-sided *Laurent expansion* of the so-called *generating function* [81, p. 14],[10]

$$e^{\frac{1}{2}\beta\left(z-\frac{1}{z}\right)} = \sum_{k=-\infty}^{\infty} J_k(\beta) z^k \tag{4.6}$$

where k is the integer *order* of the Bessel function, and β is its argument (which can be complex, but we will only consider real β). Setting $z = e^{j\omega_m t}$, where ω_m will interpreted as the *FM modulation frequency* and t as time in seconds, we obtain

$$x_m(t) \triangleq e^{j\beta\sin(\omega_m t)} = \sum_{k=-\infty}^{\infty} J_k(\beta) e^{jk\omega_m t}. \tag{4.7}$$

The last expression can be interpreted as the Fourier superposition of the sinusoidal harmonics of $\exp[j\beta\sin(\omega_m t)]$, *i.e.*, an *inverse Fourier series* sum. In other words, $J_k(\beta)$ is the amplitude of the kth harmonic in the Fourier-series expansion of the periodic signal $x_m(t)$.

Note that $J_k(\beta)$ is real when β is real. This can be seen by viewing Eq. (4.6) as the product of the series expansion for $\exp[(\beta/2)z]$ times that for $\exp[-(\beta/2)/z]$ (see footnote pertaining to Eq. (4.6)).

Figure 4.15 illustrates the first eleven Bessel functions of the first kind for arguments up to $\beta = 30$. It can be seen in the figure that when the FM index β is zero, $J_0(0) = 1$ and $J_k(0) = 0$ for all $k > 0$. Since $J_0(\beta)$ is the amplitude of the carrier frequency, there are no side bands when $\beta = 0$. As the FM index increases, the sidebands begin to grow while the carrier term diminishes. This is how FM synthesis produces an expanded, brighter bandwidth as the FM index is increased.

FM Spectra

Using the expansion in Eq. (4.7), it is now easy to determine the spectrum of sinusoidal FM. Eliminating scaling and phase offsets for simplicity in

[10] Existence of the Laurent expansion follows from the fact that the generating function is a product of an exponential function, $\exp(\beta z/2)$, and an exponential function inverted with respect to the unit circle, $\exp(-0.5\beta/z)$. It is readily verified by direct differentiation in the complex plane that the exponential is an *entire function* of z (analytic at all finite points in the complex plane) [13], and therefore the inverted exponential is analytic everywhere except at $z = 0$. The desired Laurent expansion may be obtained, in principle, by multiplying one-sided series for the exponential and inverted exponential together. The exponential series has the well known form $\exp(z) = 1 + z + z^2/2! + z^3/3! + \cdots$. The series for the inverted exponential can be obtained by inverting again ($z \leftarrow 1/z$), obtaining the appropriate exponential series, and inverting each term.

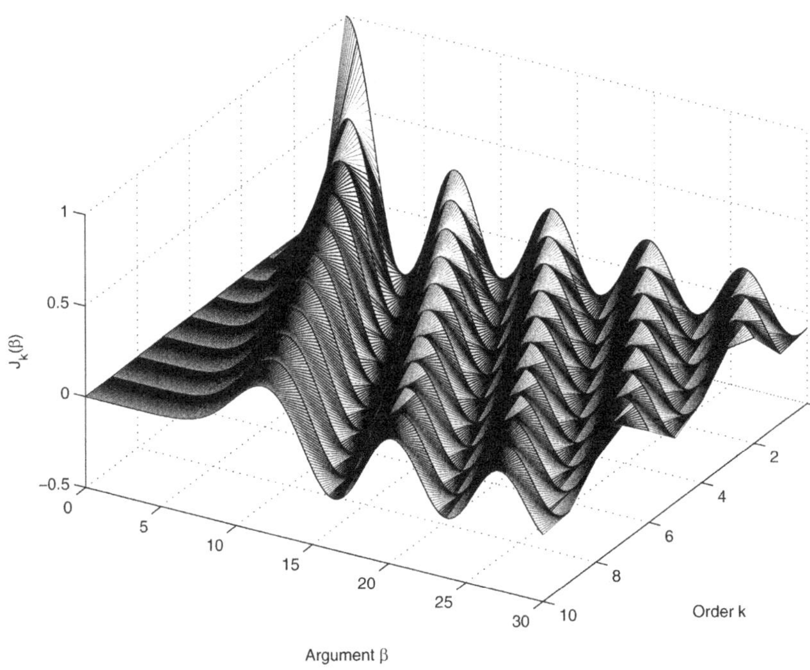

Figure 4.15: Bessel functions of the first kind for a range of orders k and argument β.

Eq. (4.5) yields

$$x(t) = \cos[\omega_c t + \beta \sin(\omega_m t)], \tag{4.8}$$

where we have changed the modulator amplitude A_m to the more traditional symbol β, called the *FM index* in FM sound synthesis contexts. Using *phasor analysis* (where *phasors* are defined below in §4.3.11 on page 59),[11] *i.e.*, expressing a real-valued FM signal as the real part of a more analytically tractable complex-valued FM signal, we obtain

$$\begin{aligned}
x(t) \triangleq \cos[\omega_c t + \beta \sin(\omega_m t)] &= \text{re}\left\{e^{j[\omega_c t + \beta \sin(\omega_m t)]}\right\} \\
&= \text{re}\left\{e^{j\omega_c t} e^{j\beta \sin(\omega_m t)}\right\} \\
&= \text{re}\left\{e^{j\omega_c t} \sum_{k=-\infty}^{\infty} J_k(\beta) e^{jk\omega_m t}\right\} \\
&= \text{re}\left\{\sum_{k=-\infty}^{\infty} J_k(\beta) e^{j(\omega_c + k\omega_m)t}\right\} \\
&= \sum_{k=-\infty}^{\infty} J_k(\beta) \cos[(\omega_c + k\omega_m)t] \qquad (4.9)
\end{aligned}$$

where we used the fact that $J_k(\beta)$ is real when β is real. We can now see clearly that the sinusoidal FM spectrum consists of an infinite number of side-bands about the carrier frequency ω_c (when $\beta \neq 0$). The side bands occur at multiples of the modulating frequency ω_m away from the carrier frequency ω_c.

4.3.7 Analytic Signals and Hilbert Transform Filters

A signal which has no negative-frequency components is called an *analytic signal*.[12] Therefore, in continuous time, every analytic signal $z(t)$ can be represented as

$$z(t) = \frac{1}{2\pi} \int_0^{\infty} Z(\omega) e^{j\omega t} d\omega$$

[11] Another example of phasor analysis can be found at

`http://ccrma.stanford.edu/~jos/filters/Phasor_Analysis.html`

(from the next book in this music signal processing series).

[12] In complex variables, a function is "analytic" at a point $z_0 \in \mathbf{C}$ if it is differentiable of all orders at each point in some neighborhood of z_0 [13]. Therefore, one might expect an "analytic signal" to be any signal which is differentiable of all orders at any point in time, *i.e.*, one that admits a fully valid Taylor expansion about any point in time. However, *all* bandlimited signals (being sums of finite-frequency sinusoids) are analytic in the complex-variables sense at every point in time. Therefore, the signal processing term "analytic signal" refers instead to a signal having "no negative frequencies". Equivalently, one could say that the spectrum of an analytic signal is "causal in the frequency domain".

where $Z(\omega)$ is the complex coefficient (setting the amplitude and phase) of the positive-frequency complex sinusoid $\exp(j\omega t)$ at frequency ω.

Any real sinusoid $A\cos(\omega t+\phi)$ may be converted to a positive-frequency complex sinusoid $A\exp[j(\omega t+\phi)]$ by simply generating a phase-quadrature component $A\sin(\omega t+\phi)$ to serve as the "imaginary part":

$$Ae^{j(\omega t+\phi)} = A\cos(\omega t+\phi) + jA\sin(\omega t+\phi)$$

The phase-quadrature component can be generated from the in-phase component by a simple quarter-cycle time shift.[13]

For more complicated signals which are expressible as a sum of many sinusoids, a *filter* can be constructed which shifts each sinusoidal component by a quarter cycle. This is called a *Hilbert transform filter*. Let $\mathcal{H}_t\{x\}$ denote the output at time t of the Hilbert-transform filter applied to the signal x. Ideally, this filter has magnitude 1 at all frequencies and introduces a phase shift of $-\pi/2$ at each positive frequency and $+\pi/2$ at each negative frequency. When a real signal $x(t)$ and its Hilbert transform $y(t) = \mathcal{H}_t\{x\}$ are used to form a new complex signal $z(t) = x(t) + jy(t)$, the signal $z(t)$ is the (complex) *analytic signal* corresponding to the real signal $x(t)$. In other words, for any real signal $x(t)$, the corresponding analytic signal $z(t) = x(t) + j\mathcal{H}_t\{x\}$ has the property that all "negative frequencies" of $x(t)$ have been "filtered out."

To see how this works, recall that these phase shifts can be impressed on a complex sinusoid by multiplying it by $\exp(\pm j\pi/2) = \pm j$. Consider the positive and negative frequency components at the particular frequency ω_0:

$$\begin{aligned} x_+(t) &\isdef e^{j\omega_0 t} \\ x_-(t) &\isdef e^{-j\omega_0 t} \end{aligned}$$

Now let's apply a -90 degrees phase shift to the positive-frequency component, and a $+90$ degrees phase shift to the negative-frequency component:

$$\begin{aligned} y_+(t) &= e^{-j\pi/2}e^{j\omega_0 t} = -je^{j\omega_0 t} \\ y_-(t) &= e^{j\pi/2}e^{-j\omega_0 t} = je^{-j\omega_0 t} \end{aligned}$$

[13] This operation is actually used in some real-world AM and FM radio receivers (particularly in digital radio receivers). The signal comes in centered about a high "carrier frequency" (such as 101 MHz for radio station FM 101), so it looks very much like a sinusoid at frequency 101 MHz. (The frequency modulation only varies the carrier frequency in a relatively tiny interval about 101 MHz. The total FM bandwidth including all the FM "sidebands" is about 100 kHz. AM bands are only 10kHz wide.) By delaying the signal by 1/4 cycle, a good approximation to the imaginary part of the analytic signal is created, and its instantaneous amplitude and frequency are then simple to compute from the analytic signal.

Adding them together gives

$$\begin{aligned} z_+(t) &\triangleq x_+(t) + jy_+(t) = e^{j\omega_0 t} - j^2 e^{j\omega_0 t} = 2e^{j\omega_0 t} \\ z_-(t) &\triangleq x_-(t) + jy_-(t) = e^{-j\omega_0 t} + j^2 e^{-j\omega_0 t} = 0 \end{aligned}$$

and sure enough, the negative frequency component is filtered out. (There is also a gain of 2 at positive frequencies.)

For a concrete example, let's start with the real sinusoid

$$x(t) = 2\cos(\omega_0 t) = e^{j\omega_0 t} + e^{-j\omega_0 t}.$$

Applying the ideal phase shifts, the Hilbert transform is

$$\begin{aligned} y(t) &= e^{j\omega_0 t - j\pi/2} + e^{-j\omega_0 t + j\pi/2} \\ &= -je^{j\omega_0 t} + je^{-j\omega_0 t} = 2\sin(\omega_0 t). \end{aligned}$$

The analytic signal is then

$$z(t) = x(t) + jy(t) = 2\cos(\omega_0 t) + j2\sin(\omega_0 t) = 2e^{j\omega_0 t},$$

by Euler's identity. Thus, in the sum $x(t) + jy(t)$, the negative-frequency components of $x(t)$ and $jy(t)$ cancel out, leaving only the positive-frequency component. This happens for any real signal $x(t)$, not just for sinusoids as in our example.

Figure 4.16 illustrates what is going on in the frequency domain. At the top is a graph of the spectrum of the sinusoid $\cos(\omega_0 t)$ consisting of impulses at frequencies $\omega = \pm\omega_0$ and zero at all other frequencies (since $\cos(\omega_0 t) = (1/2)\exp(j\omega_0 t) + (1/2)\exp(-j\omega_0 t)$). Each impulse amplitude is equal to $1/2$. (The amplitude of an impulse is its algebraic area.) Similarly, since $\sin(\omega_0 t) = (1/2j)\exp(j\omega_0 t) - (1/2j)\exp(-j\omega_0 t) = -(j/2)\exp(j\omega_0 t) + (j/2)\exp(-j\omega_0 t)$, the spectrum of $\sin(\omega_0 t)$ is an impulse of amplitude $-j/2$ at $\omega = \omega_0$ and amplitude $+j/2$ at $\omega = -\omega_0$. Multiplying $y(t)$ by j results in $j\sin(\omega_0 t) = (1/2)\exp(j\omega_0 t) - (1/2)\exp(-j\omega_0 t)$ which is shown in the third plot, Fig. 4.16c. Finally, adding together the first and third plots, corresponding to $z(t) = x(t) + jy(t)$, we see that the two positive-frequency impulses *add in phase* to give a unit impulse (corresponding to $\exp(j\omega_0 t)$), and at frequency $-\omega_0$, the two impulses, having opposite sign, *cancel* in the sum, thus creating an analytic signal z, as shown in Fig. 4.16d. This sequence of operations illustrates how the negative-frequency component $\exp(-j\omega_0 t)$ gets *filtered out* by summing $\cos(\omega_0 t)$ with $j\sin(\omega_0 t)$ to produce the analytic signal $\exp(j\omega_0 t)$ corresponding to the real signal $\cos(\omega_0 t)$.

As a final example (and application), let $x(t) = A(t)\cos(\omega t)$, where $A(t)$ is a slowly varying amplitude envelope (slow compared with ω). This

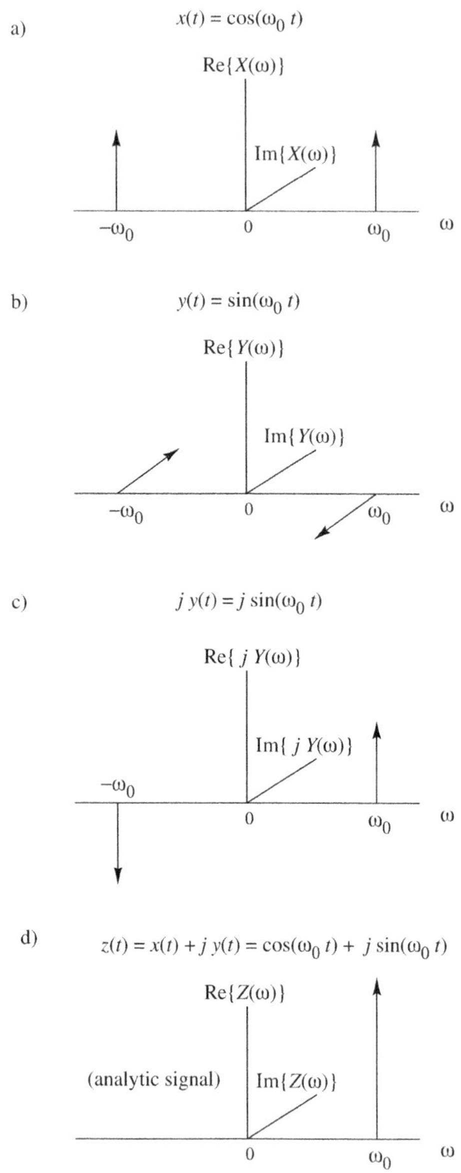

Figure 4.16: Creation of the analytic signal $z(t) = e^{j\omega_0 t}$ from the real sinusoid $x(t) = \cos(\omega_0 t)$ and the derived phase-quadrature sinusoid $y(t) = \sin(\omega_0 t)$, viewed in the frequency domain. a) Spectrum of x. b) Spectrum of y. c) Spectrum of jy. d) Spectrum of $z = x + jy$.

is an example of *amplitude modulation* applied to a sinusoid at "carrier frequency" ω (which is where you tune your AM radio). The Hilbert transform is very close to $y(t) \approx A(t)\sin(\omega t)$ (if $A(t)$ were constant, this would be exact), and the analytic signal is $z(t) \approx A(t)e^{j\omega t}$. Note that AM *demodulation*[14] is now nothing more than the *absolute value*. *I.e.*, $A(t) = |z(t)|$. Due to this simplicity, Hilbert transforms are sometimes used in making *amplitude envelope followers* for narrowband signals (*i.e.*, signals with all energy centered about a single "carrier" frequency). AM demodulation is one application of a narrowband envelope follower.

4.3.8 Generalized Complex Sinusoids

We have defined sinusoids and extended the definition to include complex sinusoids. We now extend one more step by allowing for *exponential amplitude envelopes*:

$$y(t) \stackrel{\Delta}{=} \mathcal{A}e^{st}$$

where $\mathcal{A}$ and s are *complex*, and further defined as

$$\begin{aligned} \mathcal{A} &= Ae^{j\phi} \\ s &= \sigma + j\omega. \end{aligned}$$

When $\sigma = 0$, we obtain

$$y(t) \stackrel{\Delta}{=} \mathcal{A}e^{j\omega t} = Ae^{j\phi}e^{j\omega t} = Ae^{j(\omega t+\phi)}$$

which is the complex sinusoid at amplitude A, frequency ω, and phase ϕ. More generally, we have

$$\begin{aligned} y(t) &\stackrel{\Delta}{=} \mathcal{A}e^{st} \\ &\stackrel{\Delta}{=} Ae^{j\phi}e^{(\sigma+j\omega)t} \\ &= Ae^{(\sigma+j\omega)t+j\phi} \\ &= Ae^{\sigma t}e^{j(\omega t+\phi)} \\ &= Ae^{\sigma t}\left[\cos(\omega t+\phi) + j\sin(\omega t+\phi)\right]. \end{aligned}$$

Defining $\tau = -1/\sigma$, we see that the generalized complex sinusoid is just the complex sinusoid we had before with an *exponential envelope*:

$$\begin{aligned} \text{re}\{y(t)\} &= Ae^{-t/\tau}\cos(\omega t+\phi) \\ \text{im}\{y(t)\} &= Ae^{-t/\tau}\sin(\omega t+\phi) \end{aligned}$$

[14] *Demodulation* is the process of recovering the modulation signal. For amplitude modulation (AM), the modulated signal is of the form $y(t) = A(t)\cos(\omega_c t)$, where ω_c is the "carrier frequency", $A(t) = [1 + \mu x(t)] \geq 0$ is the amplitude envelope (modulation), $x(t)$ is the modulation signal we wish to recover (the audio signal being broadcast in the case of AM radio), and μ is the modulation index for AM.

4.3.9 Sampled Sinusoids

In discrete-time audio processing, such as we normally do on a computer, we work with *samples* of continuous-time signals. Let f_s denote the sampling rate in Hz. For audio, we typically have $f_s > 40$ kHz, since the audio band nominally extends to 20 kHz. For compact discs (CDs), $f_s = 44.1$ kHz, while for digital audio tape (DAT), $f_s = 48$ kHz.

Let $T \stackrel{\Delta}{=} 1/f_s$ denote the *sampling interval* in seconds. Then to convert from continuous to discrete time, we replace t by nT, where n is an integer interpreted as the *sample number.*

The sampled generalized complex sinusoid is then

$$\begin{aligned} y(nT) &\stackrel{\Delta}{=} \mathcal{A}\, e^{st}\Big|_{t=nT} \\ &= \mathcal{A}e^{snT} = \mathcal{A}\left[e^{sT}\right]^n \\ &\stackrel{\Delta}{=} Ae^{j\phi}e^{(\sigma+j\omega)nT} \\ &= Ae^{\sigma nT}\left[\cos(\omega nT+\phi)+j\sin(\omega nT+\phi)\right] \\ &= A\left[e^{\sigma T}\right]^n\left[\cos(\omega nT+\phi)+j\sin(\omega nT+\phi)\right]. \end{aligned}$$

Thus, the sampled case consists of a sampled complex sinusoid multiplied by a sampled exponential envelope $\left[e^{\sigma T}\right]^n = e^{-nT/\tau}$.

4.3.10 Powers of z

Choose any two complex numbers z_0 and z_1, and form the sequence

$$x(n) \stackrel{\Delta}{=} z_0 z_1^n, \quad n = 0, 1, 2, 3, \ldots. \tag{4.10}$$

What are the properties of this signal? Writing the complex numbers as

$$\begin{aligned} z_0 &= Ae^{j\phi} \\ z_1 &= e^{sT} = e^{(\sigma+j\omega)T}, \end{aligned}$$

we see that the signal $x(n)$ is always a discrete-time generalized (exponentially enveloped) complex sinusoid:

$$x(n) = Ae^{\sigma nT}e^{j(\omega nT+\phi)}$$

Figure 4.17 shows a plot of a generalized (exponentially decaying, $\sigma < 0$) complex sinusoid versus time.

Note that the left projection (onto the z plane) is a decaying spiral, the lower projection (real-part vs. time) is an exponentially decaying cosine, and the upper projection (imaginary-part vs. time) is an exponentially enveloped sine wave.

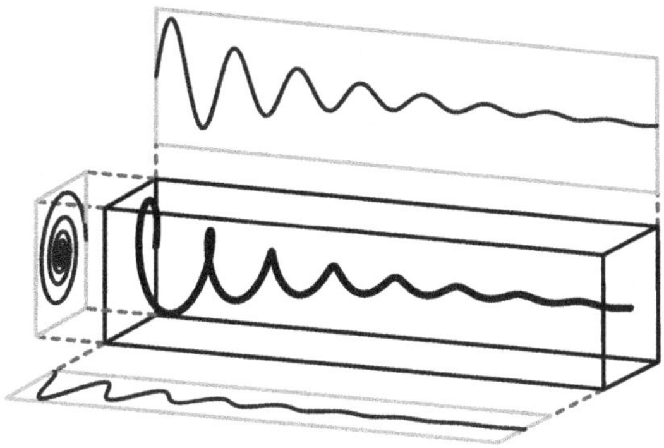

Figure 4.17: Exponentially decaying complex sinusoid and projections.

4.3.11 Phasor and Carrier Components of Sinusoids

If we restrict z_1 in Eq. (4.10) to have unit modulus, then $\sigma = 0$ and we obtain a discrete-time *complex sinusoid.*

$$x(n) \stackrel{\Delta}{=} z_0 z_1^n = \left(Ae^{j\phi}\right) e^{j\omega nT} = Ae^{j(\omega nT+\phi)}, \quad n = 0, 1, 2, 3, \ldots \tag{4.11}$$

where we have defined

$$\begin{aligned} z_0 &\stackrel{\Delta}{=} Ae^{j\phi}, \quad \text{and} \\ z_1 &\stackrel{\Delta}{=} e^{j\omega T}. \end{aligned}$$

Phasor

It is common terminology to call $z_0 = Ae^{j\phi}$ the complex sinusoid's *phasor*, and $z_1^n = e^{j\omega nT}$ its *carrier wave*.

For a *real* sinusoid,

$$x_r(n) \stackrel{\Delta}{=} A\cos(\omega nT + \phi),$$

the phasor is again defined as $z_0 = Ae^{j\phi}$ and the carrier is $z_1^n = e^{j\omega nT}$. However, in this case, the real sinusoid is recovered from its complex-sinusoid counterpart by taking the real part:

$$x_r(n) = \text{re}\left\{z_0 z_1^n\right\}$$

The *phasor magnitude* $|z_0| = A$ is the *amplitude* of the sinusoid. The *phasor angle* $\angle z_0 = \phi$ is the *phase* of the sinusoid.

When working with complex sinusoids, as in Eq. (4.11), the phasor representation $Ae^{j\phi}$ of a sinusoid can be thought of as simply the *complex amplitude* of the sinusoid. *I.e.*, it is the complex constant that multiplies the carrier term $e^{j\omega nT}$.

Why Phasors are Important

Linear, time-invariant (LTI) systems can be said to perform only four operations on a signal: copying, scaling, delaying, and adding. As a result, each output is always a *linear combination* of delayed copies of the input signal(s). (A *linear combination* is simply a weighted sum, as discussed in §5.6 on page 73.) In any linear combination of delayed copies of a complex sinusoid

$$y(n) = \sum_{i=1}^{N} g_i x(n - d_i)$$

where g_i is a weighting factor, d_i is the ith delay, and

$$x(n) \triangleq e^{j\omega nT}$$

is a complex sinusoid, the "carrier term" $e^{j\omega nT}$ can be "factored out" of the linear combination:

$$\begin{aligned} y(n) &= \sum_{i=1}^{N} g_i e^{j[\omega(n-d_i)T]} = \sum_{i=1}^{N} g_i e^{j\omega nT} e^{-j\omega d_i T} \\ &= e^{j\omega nT} \sum_{i=1}^{N} g_i e^{-j\omega d_i T} = x(n) \sum_{i=1}^{N} g_i e^{-j\omega d_i T} \end{aligned}$$

The operation of the LTI system on a complex sinusoid is thus reduced to a calculation involving only phasors, which are simply complex numbers.

Since every signal can be expressed as a linear combination of complex sinusoids, this analysis can be applied to any signal by expanding the signal into its weighted sum of complex sinusoids (*i.e.*, by expressing it as an inverse Fourier transform).

4.3.12 Importance of Generalized Complex Sinusoids

As a preview of things to come, note that one signal $y(\cdot)$[15] is *projected* onto another signal $x(\cdot)$ using an *inner product*. The inner product $\langle y, x \rangle$

[15] The notation $y(n)$ denotes a single *sample* of the signal y at sample n, while the notation $y(\cdot)$ or simply y denotes the *entire signal* for all time.

computes the *coefficient of projection*[16] of y onto x. If $x_k(n) = e^{j\omega_k nT}, n = 0, 1, 2, \ldots, N-1$ (a sampled, unit-amplitude, zero-phase, complex sinusoid), then the inner product computes the *Discrete Fourier Transform* (DFT), provided the frequencies are chosen to be $\omega_k = 2\pi k f_s/N$. For the DFT, the inner product is specifically

$$\langle y, x_k \rangle \triangleq \sum_{n=0}^{N-1} y(n)\overline{x_k(n)} = \sum_{n=0}^{N-1} y(n)e^{-j2\pi nk/N} \triangleq \text{DFT}_k(y) \triangleq Y(\omega_k).$$

Another case of importance is the *Discrete Time Fourier Transform* (DTFT), which is like the DFT except that the transform accepts an infinite number of samples instead of only N. In this case, frequency is continuous, and

$$\langle y, x_\omega \rangle = \sum_{n=0}^{\infty} y(n)e^{-j\omega nT} \triangleq \text{DTFT}_\omega(y).$$

The DTFT is what you get in the limit as the number of samples in the DFT approaches infinity. The lower limit of summation remains zero because we are assuming all signals are zero for negative time (such signals are said to be *causal*). This means we are working with *unilateral* Fourier transforms. There are also corresponding *bilateral* transforms for which the lower summation limit is $-\infty$. The DTFT is discussed further in §B.1 on page 208.

If, more generally, $x(n) = z^n$ (a sampled complex sinusoid with exponential growth or decay), then the inner product becomes

$$\langle y, x \rangle = \sum_{n=0}^{\infty} y(n)z^{-n}$$

and this is the definition of the *z transform*. It is a generalization of the DTFT: The DTFT equals the z transform evaluated on the *unit circle* in the z plane. In principle, the z transform can also be recovered from the DTFT by means of "analytic continuation" from the unit circle to the entire z plane (subject to mathematical disclaimers which are unnecessary in practical applications since they are always finite).

Why have a z transform when it seems to contain no more information than the DTFT? It is useful to generalize from the unit circle (where the DFT and DTFT live) to the entire complex plane (the z transform's domain) for a number of reasons. First, it allows transformation of *growing* functions of time such as growing exponentials; the only limitation on

[16] The coefficient of projection of a signal y onto another signal x can be thought of as a measure of how much of x is present in y. We will consider this topic in some detail later on.

growth is that it cannot be faster than exponential. Secondly, the z transform has a deeper algebraic structure over the complex plane as a whole than it does only over the unit circle. For example, the z transform of any finite signal is simply a *polynomial* in z. As such, it can be fully characterized (up to a constant scale factor) by its *zeros* in the z plane. Similarly, the z transform of an *exponential* can be characterized to within a scale factor by a single point in the z plane (the point which *generates* the exponential); since the z transform goes to infinity at that point, it is called a *pole* of the transform. More generally, the z transform of any *generalized complex sinusoid* is simply a *pole* located at the point which generates the sinusoid. Poles and zeros are used extensively in the analysis of *recursive digital filters*. On the most general level, every finite-order, linear, time-invariant, discrete-time system is fully specified (up to a scale factor) by its poles and zeros in the z plane. This topic will be taken up in detail in Book II [65].

In the *continuous-time* case, we have the *Fourier transform* which projects y onto the continuous-time sinusoids defined by $x(t) = e^{j\omega t}$, and the appropriate inner product is

$$\langle y, x\rangle = \int_0^\infty y(t)e^{-j\omega t}dt \stackrel{\Delta}{=} Y(\omega).$$

Finally, the *Laplace transform* is the continuous-time counterpart of the z transform, and it projects signals onto exponentially growing or decaying complex sinusoids:

$$\langle y, x\rangle = \int_0^\infty y(t)e^{-st}dt \stackrel{\Delta}{=} Y(s)$$

The Fourier transform equals the Laplace transform evaluated along the "$j\omega$ axis" in the s plane, *i.e.*, along the line $s = j\omega$, for which $\sigma = 0$. Also, the Laplace transform is obtainable from the Fourier transform via analytic continuation. The usefulness of the Laplace transform relative to the Fourier transform is exactly analogous to that of the z transform outlined above.

4.3.13 Comparing Analog and Digital Complex Planes

In signal processing, it is customary to use s as the Laplace transform variable for continuous-time analysis, and z as the z-transform variable for discrete-time analysis. In other words, for continuous-time systems, the frequency domain is the "s plane", while for discrete-time systems, the frequency domain is the "z plane." However, both are simply complex planes.

Figure 4.18 illustrates the various sinusoids e^{st} represented by points in the s plane. The frequency axis is $s = j\omega$, called the "$j\omega$ axis," and points

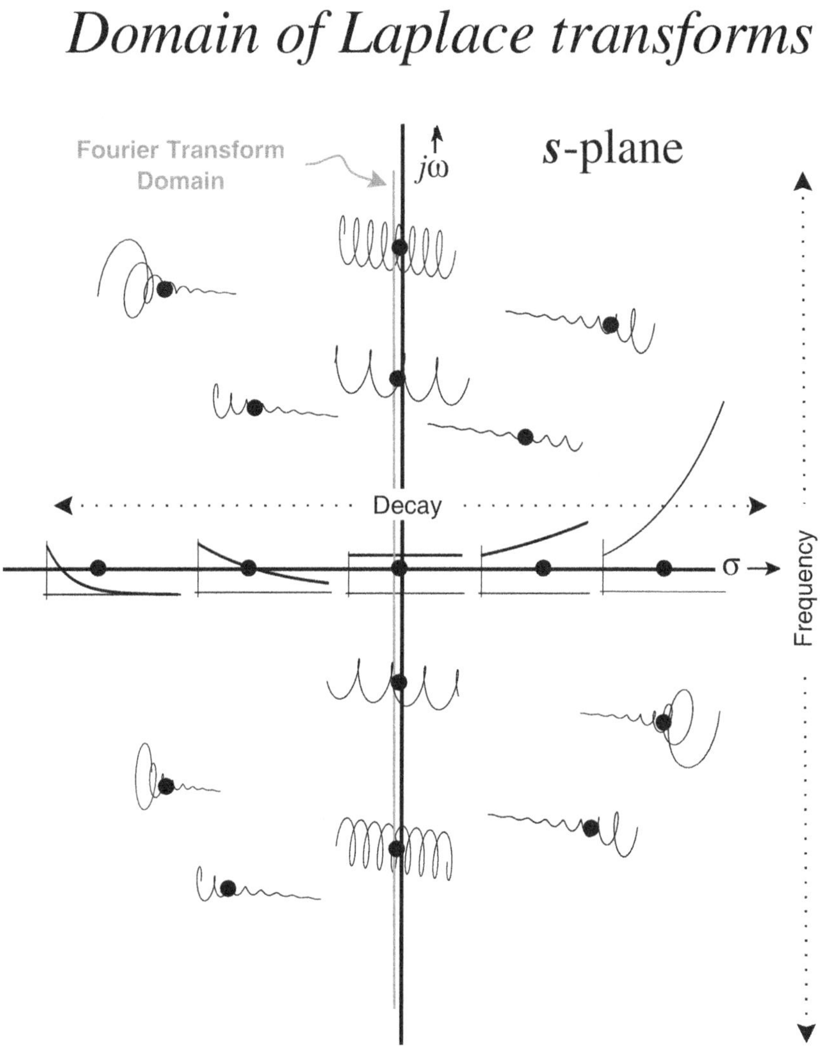

Figure 4.18: Generalized complex sinusoids represented by points in the s plane.

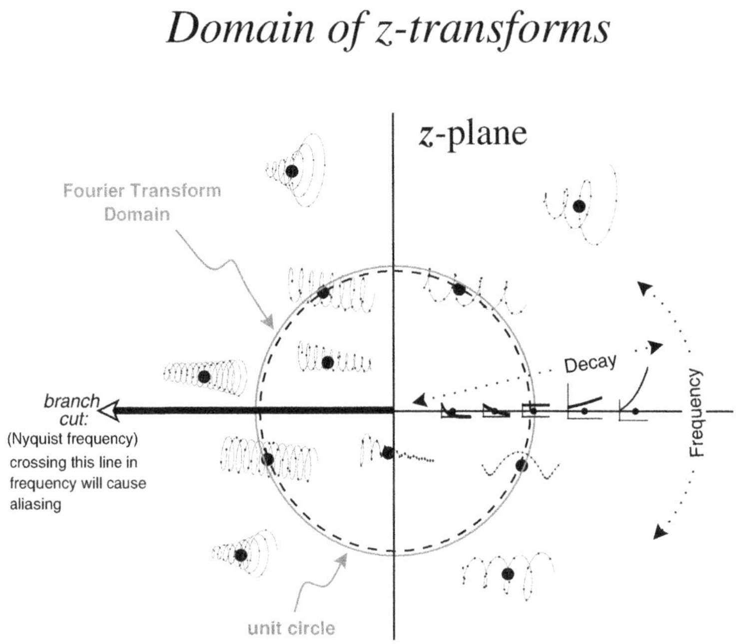

Figure 4.19: Sampled generalized complex sinusoids represented by points in the z plane.

along it correspond to complex sinusoids, with dc at $s = 0$ ($e^{j0t} \equiv 1$). The upper-half plane corresponds to positive frequencies (counterclockwise circular or corkscrew motion) while the lower-half plane corresponds to negative frequencies (clockwise motion). In the left-half plane we have decaying (stable) exponential envelopes, while in the right-half plane we have growing (unstable) exponential envelopes. Along the real axis ($s = \sigma$), we have pure exponentials. Every point in the s plane corresponds to a generalized complex sinusoid, $x(t) = \mathcal{A}e^{st}, t \geq 0$, with special cases including complex sinusoids $\mathcal{A}e^{j\omega t}$, real exponentials $Ae^{\sigma t}$, and the constant function $x(t) = 1$ (dc).

Figure 4.19 shows examples of various sinusoids $z^n = (e^{sT})^n$ represented by points in the z plane. The frequency axis is the "unit circle" $z = e^{j\omega T}$, and points along it correspond to *sampled* complex sinusoids, with dc at $z = 1$ ($1^n = [e^{j0T}]^n = 1$). As in the s plane, the upper-half plane corresponds to positive frequencies while the lower-half plane corresponds to negative frequencies. Inside the unit circle, we have decaying (stable) exponential envelopes, while outside the unit circle, we have growing (unstable) exponential envelopes. Along the positive real axis ($\text{re}\{z\} > 0,\ \text{im}\{z\} = 0$), we have pure exponentials, but along the negative real axis ($\text{re}\{z\} < 0,\ \text{im}\{z\} = 0$), we have exponentially enveloped sampled sinusoids at frequency $f_s/2$ (exponentially enveloped alternating sequences). The negative real axis in the z plane is normally a place where all signal z transforms should be zero, and all system responses should be highly attenuated, since there should never be any energy at exactly half the sampling rate (where amplitude and phase are ambiguously linked). Every point in the z plane can be said to correspond to sampled generalized complex sinusoids of the form $x(n) = \mathcal{A}z^n = \mathcal{A}[e^{sT}]^n, n \geq 0$, with special cases being sampled complex sinusoids $\mathcal{A}e^{j\omega nT}$, sampled real exponentials $Ae^{\sigma nT}$, and the constant sequence $x = [1, 1, 1, \ldots]$ (dc).

In summary, the exponentially enveloped ("generalized") complex sinusoid is the fundamental signal upon which other signals are "projected" in order to compute a Laplace transform in the continuous-time case, or a z transform in the discrete-time case. As a special case, if the exponential envelope is eliminated (set to 1), leaving only a complex sinusoid, then the projection reduces to the Fourier transform in the continuous-time case, and either the DFT (finite length) or DTFT (infinite length) in the discrete-time case. Finally, there are still other variations, such as short-time Fourier transforms (STFT) and wavelet transforms, which utilize further modifications such as projecting onto *windowed* complex sinusoids.

4.4 Sinusoid Problems

1. Figure 4.20 shows a plot of a sinusoid. From the plot, determine values for the amplitude A, phase ϕ and frequency ω_0 needed in the representation:

$$x(t) = A\cos(\omega_0 t + \phi)$$

Give the answer as numerical values, including the units when applicable.

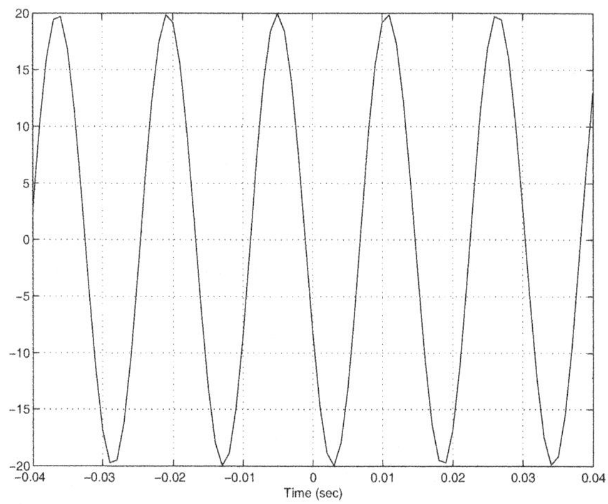

Figure 4.20: Graph of a sinusoid.

2. Given

$$x(t) = 4\sqrt{2}\sin(\omega_0 t + 45^\circ) + 3\cos(\omega_0 t)$$

   (a) Express $x(t)$ in the form $x(t) = A\cos(\omega_0 t + \phi)$, where ϕ is in radians.

   (b) Assume that $\omega_0 = 10\pi$. Make a plot of $x(t)$ over the range $-1 \leq t \leq 1$. How many periods are included in the plot?

   (c) Find a complex valued signal $\tilde{x}$ such that $x(t) = \text{re}\{\tilde{x}(t)\}$.

3. Find the amplitude and phase of the following sinusoids[17]

   (a) $\cos(\omega t) + \sin(\omega t)$

   (b) $\sin(\omega t) + \sqrt{3}\cos(\omega t)$

   [Hint: Obtain the form $A\sin(\omega t + \phi)$ and report A and ϕ.]

[17] Recall the trigonometric identity $\sin(A + B) = \sin(A)\cos(B) + \sin(B)\cos(A)$.

4. Define the discrete-time generalized sinusoid

$$x[n] = X z_0^n, \quad n = 0, 1, 2, \ldots,$$

where

$$\begin{aligned} X &= 2e^{j\pi/4}, \\ z_0 &= 0.9e^{j\pi/8}. \end{aligned}$$

   (a) What is the amplitude of this sinusoid? What is the phase in radians? What is the phase in cycles? What is the phase in degrees?
   (b) Plot $\text{re}\{X z_0^n\}$ and $\text{im}\{X z_0^n\}$ versus n.
   (c) Plot $X z_0^n$ as a collection of points in the complex plane (imaginary part versus real part).
   (d) What is the time constant τ of decay (in samples)?
   (e) What is the 60 dB decay time T_{60} in time constants?
   (f) What is T_{60} in samples?
   (g) What is the 80 dB decay time T_{80} in time constants?
   (h) What is T_{80} in samples?
   (i) If the sampling rate is 800 Hz, what are τ and T_{60} in seconds, and what is the frequency of the sinusoid in Hz?

5. For the continuous-time signal $x(t) \stackrel{\Delta}{=} 5\cos(400\pi t + \pi/4)$, what is the output signal $y(t)$ obtained when $x(t)$ is passed through an ideal filter which passes all positive-frequency components with a gain of 2, and which rejects all negative-frequency components (with a gain of 0)?

Chapter 5

Geometric Signal Theory

This chapter provides an introduction to the elements of geometric signal theory, including vector spaces, norms, inner products, orthogonality, projection of one signal onto another, and elementary vector space operations. First, however, we will "get our bearings" with respect to the DFT.

5.1 The DFT

For a length N complex sequence $x(n)$, $n = 0, 1, 2, \ldots, N-1$, the *discrete Fourier transform* (DFT) is defined by

$$X(\omega_k) \triangleq \sum_{n=0}^{N-1} x(n) e^{-j\omega_k t_n} = \sum_{n=0}^{N-1} x(n) e^{-j2\pi kn/N}, \quad k = 0, 1, 2, \ldots N-1$$

$$\begin{aligned} t_n &\triangleq nT = n\text{th sampling instant (sec)} \\ \omega_k &\triangleq k\Omega = k\text{th frequency sample (rad/sec)} \\ T &\triangleq 1/f_s = \text{time sampling interval (sec)} \\ \Omega &\triangleq 2\pi f_s/N = \text{frequency sampling interval (rad/sec)} \end{aligned}$$

We are now in a position to have a full understanding of the transform *kernel*:

$$e^{-j\omega_k t_n} = \cos(\omega_k t_n) - j\sin(\omega_k t_n)$$

The kernel consists of samples of a complex sinusoid at N discrete frequencies ω_k uniformly spaced between 0 and the sampling rate $\omega_s \triangleq 2\pi f_s$. All that remains is to understand the purpose and function of the summation over n of the pointwise product of $x(n)$ times each complex sinusoid. We will learn that this can be interpreted as an *inner product* operation which computes the *coefficient of projection* of the signal x onto the complex sinusoid $\cos(\omega_k t_n) + j\sin(\omega_k t_n)$. As such, $X(\omega_k)$, the DFT at frequency ω_k, is a measure of the amplitude and phase of the complex sinusoid which is present in the input signal x at that frequency. This is the basic function of all linear transform summations (in discrete time) and integrals (in continuous time) and their kernels.

5.2 Signals as Vectors

For the DFT, all signals and spectra are length N. A length N sequence x can be denoted by $x(n)$, $n = 0, 1, 2, \ldots N-1$, where $x(n)$ may be real ($x \in \mathbf{R}^N$) or complex ($x \in \mathbf{C}^N$). We now wish to regard x as a *vector*[1] $\underline{x}$ in an N dimensional *vector space.* That is, each sample $x(n)$ is regarded as a *coordinate* in that space. A *vector* $\underline{x}$ is mathematically a single *point* in N-space represented by a list of coordinates $(x_0, x_1, x_2, \ldots, x_{N-1})$ called an N*-tuple.* (The notation x_n means the same thing as $x(n)$.) It can be interpreted geometrically as an arrow in N-space from the origin $\underline{0} \triangleq (0, 0, \ldots, 0)$ to the point $\underline{x} \triangleq (x_0, x_1, x_2, \ldots, x_{N-1})$.

We define the following as equivalent:

$$x \triangleq \underline{x} \triangleq x(\cdot) \triangleq (x_0, x_1, \ldots, x_{N-1}) \triangleq [x_0, x_1, \ldots, x_{N-1}] \triangleq [x_0 \; x_1 \; \cdots \; x_{N-1}]$$

where $x_n \triangleq x(n)$ is the nth sample of the signal (vector) x. From now on, unless specifically mentioned otherwise, *all signals are length* N.

The reader comfortable with vectors, vector addition, and vector subtraction may skip to §5.6 on page 73.

An Example Vector View: $N = 2$

Consider the example two-sample signal $x = (2, 3)$ graphed in Fig. 5.1.

Under the geometric interpretation of a length N signal, each sample is a *coordinate* in the N dimensional space. Signals which are only two samples long are not terribly interesting to hear,[2] but they are easy to plot geometrically.

[1] We'll use an underline to emphasize the vector interpretation, but there is no difference between x and $\underline{x}$. For purposes of this book, a *signal* is the same thing as a *vector.*

[2] Actually, two-sample signals with variable amplitude and spacing between the samples provide very interesting tests of pitch perception, especially when the samples have opposite sign [53].

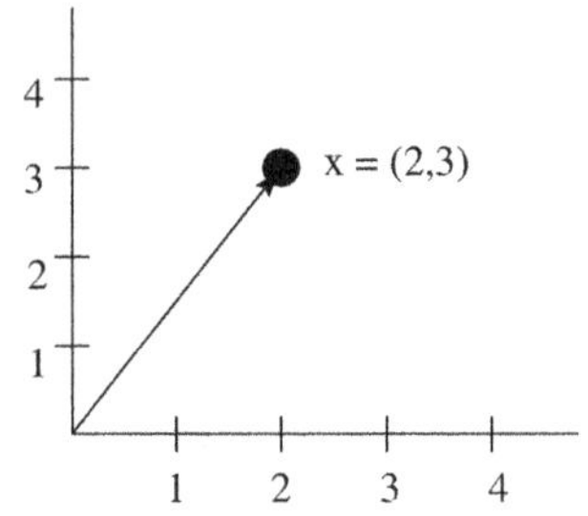

Figure 5.1: A length 2 signal $x = (2, 3)$ plotted as a vector in 2D space.

5.3 Vector Addition

Given two vectors in $\mathbf{R}^N$, say

$$\begin{aligned} \underline{x} &\stackrel{\Delta}{=} (x_0, x_1, \ldots, x_{N-1}) \\ \underline{y} &\stackrel{\Delta}{=} (y_0, y_1, \ldots, y_{N-1}), \end{aligned}$$

the *vector sum* is defined by *elementwise* addition. If we denote the sum by $\underline{w} \stackrel{\Delta}{=} \underline{x} + \underline{y}$, then we have $\underline{w}_n = x_n + y_n$ for $n = 0, 1, 2, \ldots, N-1$. We could also write $\underline{w}(n) = x(n) + y(n)$ for $n = 0, 1, 2, \ldots, N-1$ if preferred.

The vector diagram for the sum of two vectors can be found using the parallelogram rule, as shown in Fig. 5.2 for $N = 2$, $\underline{x} = (2, 3)$, and $\underline{y} = (4, 1)$.

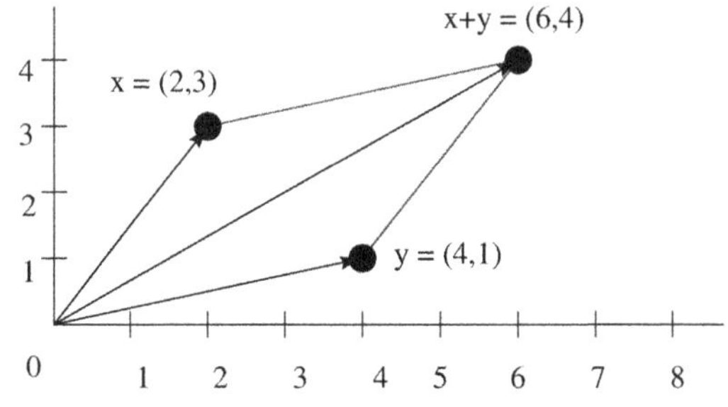

Figure 5.2: Geometric interpretation of a length 2 vector sum.

Also shown are the lighter construction lines which complete the parallelogram started by $\underline{x}$ and $\underline{y}$, indicating where the endpoint of the sum $\underline{x}+\underline{y}$

lies. Since it is a parallelogram, the two construction lines are congruent to the vectors $\underline{x}$ and $\underline{y}$. As a result, the vector sum is often expressed as a *triangle* by translating the origin of one member of the sum to the tip of the other, as shown in Fig. 5.3.

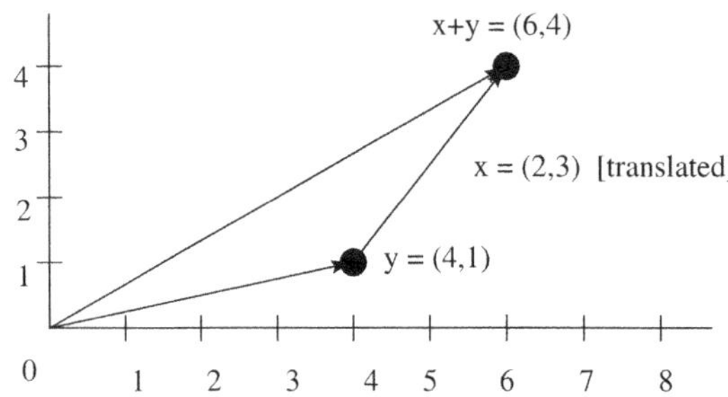

Figure 5.3: Vector sum, translating one vector to the tip of the other.

In the figure, $\underline{x}$ was translated to the tip of $\underline{y}$. This depicts $y+x$, since "x picks up where y leaves off." It is equally valid to translate $\underline{y}$ to the tip of $\underline{x}$, because vector addition is *commutative*, *i.e.*, $\underline{x} + \underline{y} = \underline{y} + \underline{x}$.

5.4 Vector Subtraction

Figure 5.4 illustrates the vector difference $\underline{w} = \underline{x} - \underline{y}$ between $\underline{x} = (2,3)$ and $\underline{y} = (4,1)$. From the coordinates, we compute $\underline{w} = \underline{x} - \underline{y} = (-2,2)$.

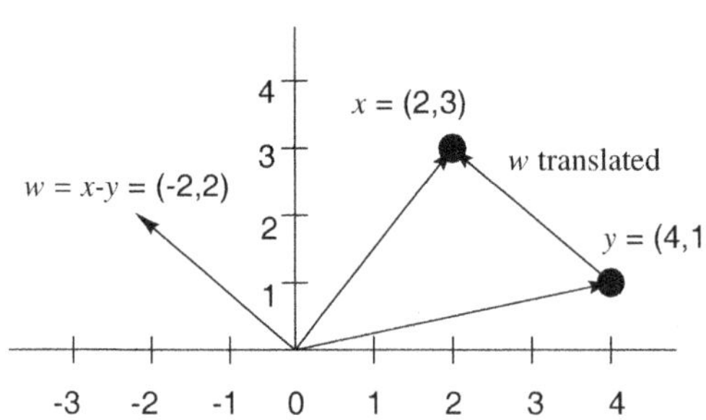

Figure 5.4: Geometric interpretation of a difference vector.

Note that the difference vector $\underline{w}$ may be drawn from the tip of $\underline{y}$ to the tip of $\underline{x}$ rather than from the origin to the point $(-2,2)$; this is a customary practice which emphasizes relationships among vectors, but the translation in the plot has no effect on the mathematical definition or properties of the vector. Subtraction, however, is not commutative.

To ascertain the proper orientation of the difference vector $\underline{w} = \underline{x} - \underline{y}$, rewrite its definition as $\underline{x} = \underline{y} + \underline{w}$, and then it is clear that the vector $\underline{x}$ should be the sum of vectors $\underline{y}$ and $\underline{w}$, hence the arrowhead is on the correct endpoint. Or remember "$x - y$ points to x," or "$x - y$ is x from y."

5.5 Scalar Multiplication

A *scalar* is any constant value used as a *scale factor* applied to a vector. Mathematically, all of our scalars will be either real or complex numbers.[3] For example, if $\underline{x} \in \mathbf{C}^N$ denotes a vector of N complex elements, and $\alpha \in \mathbf{C}$ denotes a complex scalar, then

$$\alpha\,\underline{x} \stackrel{\Delta}{=} (\alpha\,x_1, \alpha\,x_2, \ldots, \alpha\,x_N)$$

denotes the *scalar multiplication* of $\underline{x}$ by α. Thus, multiplication of a vector by a scalar is done in the obvious way, which is to multiply each coordinate of the vector by the scalar.

In signal processing, we think of scalar multiplication as applying some constant *scale factor* to a signal, *i.e.*, multiplying each sample of the signal by the same constant number. For example, a 6 dB boost can be carried out by multiplying each sample of a signal by 2, in which case 2 is the scalar. When the scalar magnitude is greater than one, it is often called a *gain factor*, and when it is less than one, an *attenuation*.

5.6 Linear Combination of Vectors

A *linear combination* of vectors is a *sum of scalar multiples* of those vectors. That is, given a set of M vectors $\underline{x}_i$ of the same type,[4] such as $\mathbf{R}^N$ (they must have the same number of elements so they can be added), a linear combination is formed by multiplying each vector by a scalar α_i and summing to produce a new vector $\underline{y}$ of the same type:

$$\underline{y} = \alpha_1\underline{x}_1 + \alpha_2\underline{x}_2 + \cdots + \alpha_M\underline{x}_M$$

---

[3]More generally, scalars are often defined as members of some *mathematical field*—usually the same field used for the vector elements (coordinates, signal samples).

[4]As we'll discuss in §5.7 on page 74 below, vectors of the "same type" are typically taken to be members of the same *vector space*.

For example, let $\underline{x}_1 = (1,2,3)$, $\underline{x}_2 = (4,5,6)$, $\alpha_1 = 2$, and $\alpha_2 = 3$. Then the linear combination of $\underline{x}_1$ and $\underline{x}_2$ is given by

$$\underline{y} = \alpha_1\underline{x}_1+\alpha_2\underline{x}_2 = 2\cdot(1,2,3)+3\cdot(4,5,6) = (2,4,6)+(12,15,18) = (14,19,24).$$

In signal processing, we think of a linear combination as a *signal mix*. Thus, the output of a *mixing console* may be regarded as a *linear combination* of the input signal tracks.

5.7 Linear Vector Space

A set of vectors may be called a *linear vector space* if it is *closed* under linear combinations. That is, given any two vectors $\underline{x}_1$ and $\underline{x}_2$ from the set, the linear combination

$$\underline{y} = \alpha_1\underline{x}_1 + \alpha_2\underline{x}_2$$

is also in the set, for all scalars α_1 and α_2. In our context, most generally, the vector coordinates and the scalars can be any complex numbers. Since complex numbers are closed under multiplication and addition, it follows that the set of all vectors in $\mathbf{C}^N$ with complex scalars ($\alpha \in \mathbf{C}$) forms a linear vector space. The same can be said of real length-N vectors in $\mathbf{R}^N$ with real scalars ($\alpha \in \mathbf{R}$). However, real vectors with complex scalars do not form a vector space, since scalar multiplication can take a real vector to a complex vector outside of the set of real vectors.

5.8 Signal Metrics

This section defines some useful functions of signals (vectors).

The *mean* of a signal x (more precisely the "sample mean") is defined as the *average value* of its samples:[5]

$$\mu_x \triangleq \frac{1}{N}\sum_{n=0}^{N-1} x_n \qquad \text{(mean of } x\text{)}$$

The *total energy* of a signal x is defined as the *sum of squared moduli*:

$$\mathcal{E}_x \triangleq \sum_{n=0}^{N-1} |x_n|^2 \qquad \text{(energy of } x\text{)}$$

[5] In this section, x denotes a signal, while in the previous sections, we used an underline ($\underline{x}$) to emphasize the vector interpretation of a signal. One might worry that it is now too easy to confuse signals (vectors) and scale factors (scalars), but this is usually not the case: signal names are generally taken from the end of the Roman alphabet (u, v, w, x, y), while scalar symbols are chosen from the beginning of the Roman $(a, b, c, \ldots)$ and Greek $(\alpha, \beta, \gamma, \ldots)$ alphabets. Also, formulas involving signals are typically specified on the sample level, so that signals are usually indexed $(x(n))$ or subscripted (x_n).

In physics, energy (the "ability to do work") and work are in units of "force times distance," "mass times velocity squared," or other equivalent combinations of units.[6] In digital signal processing, physical units are routinely discarded, and signals are renormalized whenever convenient. Therefore, $\mathcal{E}_x$ is defined above without regard for constant scale factors such as "wave impedance" or the sampling interval T.

The *average power* of a signal x is defined as the *energy per sample*:

$$\mathcal{P}_x \triangleq \frac{\mathcal{E}_x}{N} = \frac{1}{N}\sum_{n=0}^{N-1}|x_n|^2 \qquad \text{(average power of } x\text{)}$$

Another common description when x is real is the *mean square*. When x is a complex sinusoid, *i.e.*, $x(n) = Ae^{j(\omega nT+\phi)}$, then $\mathcal{P}_x = A^2$; in other words, for complex sinusoids, the average power equals the *instantaneous power* which is the amplitude squared. For real sinusoids, $y_n = \text{re}\{x_n\} = A\cos(\omega nT+\phi)$, we have $\mathcal{P}_y = A^2/2$.

Power is always in physical units of energy per unit time. It therefore makes sense to define the average signal power as the total signal energy divided by its length. We normally work with signals which are functions of time. However, if the signal happens instead to be a function of distance (*e.g.*, samples of displacement along a vibrating string), then the "power" as defined here still has the interpretation of a *spatial energy density*. Power, in contrast, is a *temporal energy density*.

The *root mean square* (RMS) level of a signal x is simply $\sqrt{\mathcal{P}_x}$. However, note that in practice (especially in audio work) an RMS level is typically computed after subtracting out any nonzero mean value.

The *variance* (more precisely the *sample variance*) of the signal x is defined as the power of the signal with its mean removed:[7]

$$\sigma_x^2 \triangleq \frac{1}{N}\sum_{n=0}^{N-1}|x_n - \mu_x|^2 \qquad \text{(sample variance of } x\text{)}$$

---

[6]The energy of a *pressure wave* is the integral over time and area of the squared pressure divided by the wave impedance the wave is traveling in. The energy of a *velocity wave* is the integral over time of the squared velocity times the wave impedance. In audio work, a signal x is typically a list of *pressure samples* derived from a microphone signal, or it might be samples of *force* from a piezoelectric transducer, *velocity* from a magnetic guitar pickup, and so on. In all of these cases, the total physical energy associated with the signal is proportional to the sum of squared signal samples. Physical connections in signal processing are explored more fully in Book III of the Music Signal Processing Series [66], (available online).

[7]For reasons beyond the scope of this book, when the sample mean μ_x is estimated as the average value of the same N samples used to compute the sample variance σ_x^2, the sum should be divided by $N-1$ rather than N to avoid a *bias* [31].

It is quick to show that, for real signals, we have

$$\sigma_x^2 = \mathcal{P}_x - \mu_x^2$$

which is the "mean square minus the mean squared." We think of the variance as the power of the non-constant signal components (*i.e.*, everything but dc). The terms "sample mean" and "sample variance" come from the field of *statistics*, particularly the theory of *stochastic processes*. The field of *statistical signal processing* [25, 31, 62] is firmly rooted in statistical topics such as "probability," "random variables," "stochastic processes," and "time series analysis." In this book, we will only touch lightly on a few elements of statistical signal processing in a self-contained way.

The *norm* (more specifically, the L^2 *norm*, or *Euclidean norm*) of a signal x is defined as the square root of its total energy:

$$\|x\| \isdef \sqrt{\mathcal{E}_x} = \sqrt{\sum_{n=0}^{N-1} |x_n|^2} \qquad \text{(norm of } x\text{)}$$

We think of $\|x\|$ as the *length* of the vector x in N-space. Furthermore, $\|x - y\|$ is regarded as the *distance* between x and y. The norm can also be thought of as the "absolute value" or "radius" of a vector.[8]

Example: Going back to our simple 2D example $x = [2, 3]$, we can compute its norm as $\|x\| = \sqrt{2^2 + 3^2} = \sqrt{13} = 3.6056\ldots$. The physical interpretation of the norm as a distance measure is shown in Fig. 5.5.

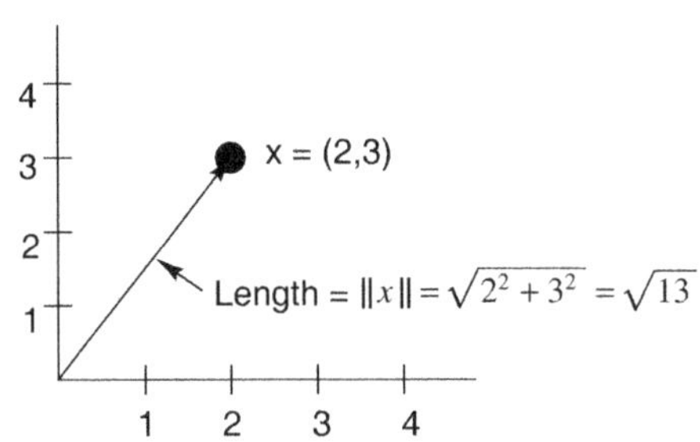

Figure 5.5: Geometric interpretation of a signal norm in 2D.

[8]You might wonder why the norm of $\underline{x}$ is not written as $|\underline{x}|$. There would be no problem with this since $|\underline{x}|$ is otherwise undefined for vectors. However, the historically adopted notation is instead $\|\underline{x}\|$.

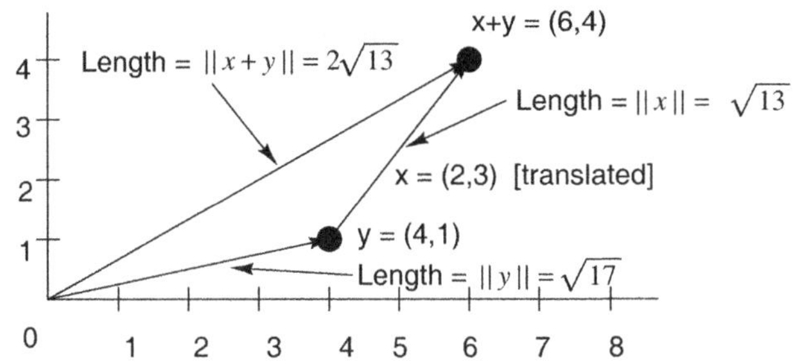

Figure 5.6: Length of vectors in sum.

Example: Let's also look again at the vector-sum example, which is redrawn in Fig. 5.6. The norm of the vector sum $w = x + y$ is

$$\|w\| \triangleq \|x + y\| \triangleq \|(2,3) + (4,1)\| = \|(6,4)\| = \sqrt{6^2 + 4^2} = \sqrt{52} = 2\sqrt{13}$$

while the norms of x and y are $\sqrt{13}$ and $\sqrt{17}$, respectively. We find that $\|x+y\| < \|x\|+\|y\|$, which is an example of the *triangle inequality*. (Equality occurs only when x and y are collinear, as can be seen geometrically from studying Fig. 5.6.)

Example: Consider the vector-difference example shown in Fig. 5.7. The norm of the difference vector $w = x - y$ is

$$\|w\| \triangleq \|x - y\| \triangleq \|(2,3) - (4,1)\| = \|(-2,2)\| = \sqrt{(-2)^2 + (2)^2} = 2\sqrt{2}.$$

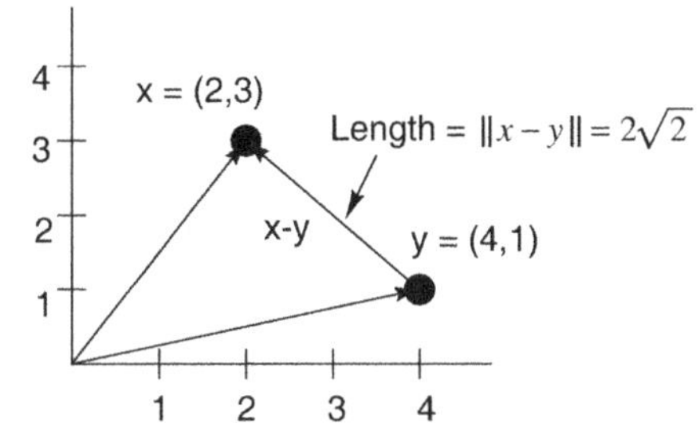

Figure 5.7: Length of a difference vector.

5.8.1 Other Lp Norms

Since our main norm is the square root of a sum of squares,

$$\|x\| \stackrel{\Delta}{=} \sqrt{\mathcal{E}_x} = \sqrt{\sum_{n=0}^{N-1} |x_n|^2} \qquad \text{(norm of } x\text{)},$$

we are using what is called an L^2 *norm* and we may write $\|x\|_2$ to emphasize this fact.

We could equally well have chosen a *normalized L^2 norm*:

$$\|x\|_{\tilde{2}} \stackrel{\Delta}{=} \sqrt{\mathcal{P}_x} = \sqrt{\frac{1}{N}\sum_{n=0}^{N-1} |x_n|^2} \qquad \text{(normalized } L^2 \text{ norm of } x\text{)}$$

which is simply the "RMS level" of x ("Root Mean Square").

More generally, the (unnormalized) L^p *norm* of $x \in \mathbf{C}^N$ is defined as

$$\|x\|_p \stackrel{\Delta}{=} \left(\sum_{n=0}^{N-1} |x_n|^p\right)^{1/p}.$$

(The normalized case would include $1/N$ in front of the summation.) The most interesting L^p norms are

- $p = 1$: The L^1, "absolute value," or "city block" norm.
- $p = 2$: The L^2, "Euclidean," "root energy," or "least squares" norm.
- $p = \infty$: The L^∞, "Chebyshev," "supremum," "minimax," or "uniform" norm.

Note that the case $p = \infty$ is a limiting case which becomes

$$\|x\|_\infty = \max_{0 \le n < N} |x_n|.$$

5.8.2 Norm Properties

There are many other possible choices of norm. To qualify as a norm on $\mathbf{C}^N$, a real-valued signal-function $f(\underline{x})$ must satisfy the following three properties:

1. $f(\underline{x}) \geq 0$, with $0 \Leftrightarrow \underline{x} = \underline{0}$
2. $f(\underline{x} + \underline{y}) \leq f(\underline{x}) + f(\underline{y})$
3. $f(c\underline{x}) = |c|\, f(\underline{x})$, $\forall c \in \mathbf{C}$

The first property, "positivity," says the norm is nonnegative, and only the zero vector has norm zero. The second property is "subadditivity" and is sometimes called the "triangle inequality" for reasons that can be seen by studying Fig. 5.6 on page 77. The third property says the norm is "absolutely homogeneous" with respect to scalar multiplication. (The scalar c can be complex, in which case the angle of c has no effect).

Banach Spaces

Mathematically, what we are working with so far is called a *Banach space*, which is a *normed* linear vector space. To summarize, we defined our vectors as any list of N real or complex numbers which we interpret as coordinates in the N-dimensional vector space. We also defined vector addition (§5.3) and scalar multiplication (§5.5) in the obvious way. To have a linear vector space (§5.7), it must be *closed* under vector addition and scalar multiplication (linear combinations). *I.e.*, given any two vectors $\underline{x} \in \mathbf{C}^N$ and $\underline{y} \in \mathbf{C}^N$ from the vector space, and given any two scalars $\alpha \in \mathbf{C}$ and $\beta \in \mathbf{C}$ from the field of scalars $\mathbf{C}^N$, the linear combination $\alpha\underline{x} + \beta\underline{y}$ must also be in the space. Since we have used the field of complex numbers $\mathbf{C}$ (or real numbers $\mathbf{R}$) to define both our scalars and our vector components, we have the necessary closure properties so that any linear combination of vectors from $\mathbf{C}^N$ lies in $\mathbf{C}^N$. Finally, the definition of a norm (any norm) elevates a vector space to a *Banach space*.

5.9 The Inner Product

The *inner product* (or "dot product", or "scalar product") is an operation on two vectors which produces a scalar. Defining an inner product for a Banach space specializes it to a *Hilbert space* (or "inner product space"). There are many examples of Hilbert spaces, but we will only need $\{\mathbf{C}^N, \mathbf{C}\}$ for this book (complex length N vectors, and complex scalars).

The *inner product* between (complex) N-vectors $\underline{u}$ and $\underline{v}$ is defined by[9]

$$\boxed{\langle \underline{u}, \underline{v} \rangle \triangleq \sum_{n=0}^{N-1} u(n)\overline{v(n)}.}$$

The complex conjugation of the second vector is done in order that a *norm* will be *induced* by the inner product:[10]

$$\langle \underline{u}, \underline{u} \rangle = \sum_{n=0}^{N-1} u(n)\overline{u(n)} = \sum_{n=0}^{N-1} |u(n)|^2 \triangleq \mathcal{E}_u = \|u\|^2$$

As a result, the inner product is *conjugate symmetric*:

$$\langle \underline{v}, \underline{u} \rangle = \overline{\langle \underline{u}, \underline{v} \rangle}$$

Note that the inner product takes $\mathbf{C}^N \times \mathbf{C}^N$ to $\mathbf{C}$. That is, two length N complex vectors are mapped to a complex scalar.

Example: For $N = 3$ we have, in general,

$$\langle \underline{u}, \underline{v} \rangle = u_0\overline{v_0} + u_1\overline{v_1} + u_2\overline{v_2}.$$

Let

$$\begin{aligned} \underline{u} &= [0, j, 1] \\ \underline{v} &= [1, j, j]. \end{aligned}$$

Then

$$\langle \underline{u}, \underline{v} \rangle = 0 \cdot 1 + j \cdot (-j) + 1 \cdot (-j) = 0 + 1 + (-j) = 1 - j.$$

See §I.3.1 on page 277 regarding computation of inner products in the matlab programming language.

[9] From some points of view, it is more elegant to conjugate the first operand in the definition of the inner product. However, for explaining the DFT, conjugating the second operand is better. The former case arises when expressing inner product $\langle \underline{u}, \underline{v} \rangle$ as a vector operation $\underline{v}^*\underline{u}$, where $\underline{v}^* \triangleq \overline{\underline{v}}^T$ denotes the *Hermitian transpose* of the vector $\underline{v}$. Either convention works out fine, but it is best to choose one and stick with it forever.

[10] Remember that a norm must be a *real*-valued function of a signal (vector).

5.9.1 Linearity of the Inner Product

Any function $f(\underline{u})$ of a vector $\underline{u} \in \mathbf{C}^N$ (which we may call an *operator* on $\mathbf{C}^N$) is said to be *linear* if for all $\underline{u} \in \mathbf{C}^N$ and $\underline{v} \in \mathbf{C}^N$, and for all scalars α and β in $\mathbf{C}$,

$$f(\alpha\underline{u} + \beta\underline{v}) = \alpha f(\underline{u}) + \beta f(\underline{v}).$$

A linear operator thus "commutes with mixing." Linearity consists of two component properties:

- *additivity*: $f(\underline{u} + \underline{v}) = f(\underline{u}) + f(\underline{v})$
- *homogeneity*: $f(\alpha\underline{u}) = \alpha f(\underline{u})$

A function of multiple vectors, *e.g.*, $f(\underline{u}, \underline{v}, \underline{w})$ can be linear or not with respect to each of its arguments.

The inner product $\langle \underline{u}, \underline{v} \rangle$ is *linear in its first argument*, *i.e.*, for all $\underline{u}, \underline{v}, \underline{w} \in \mathbf{C}^N$, and for all $\alpha, \beta \in \mathbf{C}^N$,

$$\langle \alpha\underline{u} + \beta\underline{v}, \underline{w} \rangle = \alpha \langle \underline{u}, \underline{w} \rangle + \beta \langle \underline{v}, \underline{w} \rangle .$$

This is easy to show from the definition:

$$\begin{aligned}
\langle \alpha\underline{u} + \beta\underline{v}, \underline{w} \rangle &\triangleq \sum_{n=0}^{N-1} [\alpha u(n) + \beta v(n)] \overline{w(n)} \\
&= \sum_{n=0}^{N-1} \alpha u(n)\overline{w(n)} + \sum_{n=0}^{N-1} \beta v(n)\overline{w(n)} \\
&= \alpha \sum_{n=0}^{N-1} u(n)\overline{w(n)} + \beta \sum_{n=0}^{N-1} v(n)\overline{w(n)} \\
&\triangleq \alpha \langle \underline{u}, \underline{w} \rangle + \beta \langle \underline{v}, \underline{w} \rangle
\end{aligned}$$

The inner product is also *additive* in its second argument, *i.e.*,

$$\langle \underline{u}, \underline{v} + \underline{w} \rangle = \langle \underline{u}, \underline{v} \rangle + \langle \underline{u}, \underline{w} \rangle ,$$

but it is only *conjugate homogeneous* (or *antilinear*) in its second argument, since

$$\langle \underline{u}, \alpha\underline{v} \rangle = \overline{\alpha} \langle \underline{u}, \underline{v} \rangle \neq \alpha \langle \underline{u}, \underline{v} \rangle .$$

The inner product *is* strictly linear in its second argument with respect to *real* scalars a and b:

$$\langle \underline{u}, a\underline{v} + b\underline{w} \rangle = a \langle \underline{u}, \underline{v} \rangle + b \langle \underline{u}, \underline{w} \rangle , \quad a, b \in \mathbf{R}$$

where $\underline{u}, \underline{v}, \underline{w} \in \mathbf{C}^N$.

Since the inner product is linear in both of its arguments for real scalars, it may be called a *bilinear operator* in that context.

5.9.2 Norm Induced by the Inner Product

We may define a *norm* on $\underline{u} \in \mathbf{C}^N$ using the inner product:

$$\boxed{\|\underline{u}\| \triangleq \sqrt{\langle \underline{u}, \underline{u} \rangle}}$$

It is straightforward to show that properties 1 and 3 of a norm hold (see §5.8.2 on page 79). Property 2 follows easily from the Schwarz Inequality which is derived in the following subsection. Alternatively, we can simply observe that the inner product induces the well known L^2 norm on $\mathbf{C}^N$.

5.9.3 Cauchy-Schwarz Inequality

The *Cauchy-Schwarz Inequality* (or "Schwarz Inequality") states that for all $\underline{u} \in \mathbf{C}^N$ and $\underline{v} \in \mathbf{C}^N$, we have

$$\boxed{|\langle \underline{u}, \underline{v} \rangle| \leq \|\underline{u}\| \cdot \|\underline{v}\|}$$

with equality if and only if $\underline{u} = c\underline{v}$ for some scalar c.

We can quickly show this for real vectors $\underline{u} \in \mathbf{R}^N$, $\underline{v} \in \mathbf{R}^N$, as follows: If either $\underline{u}$ or $\underline{v}$ is zero, the inequality holds (as equality). Assuming both are nonzero, let's scale them to unit-length by defining the normalized vectors $\underline{\tilde{u}} \triangleq \underline{u}/\|\underline{u}\|$, $\underline{\tilde{v}} \triangleq \underline{v}/\|\underline{v}\|$, which are unit-length vectors lying on the "unit ball" in $\mathbf{R}^N$ (a hypersphere of radius 1). We have

$$\begin{aligned}
0 \leq \|\underline{\tilde{u}} - \underline{\tilde{v}}\|^2 &= \langle \underline{\tilde{u}} - \underline{\tilde{v}}, \underline{\tilde{u}} - \underline{\tilde{v}} \rangle \\
&= \langle \underline{\tilde{u}}, \underline{\tilde{u}} - \underline{\tilde{v}} \rangle - \langle \underline{\tilde{v}}, \underline{\tilde{u}} - \underline{\tilde{v}} \rangle \\
&= \langle \underline{\tilde{u}}, \underline{\tilde{u}} \rangle - \langle \underline{\tilde{u}}, \underline{\tilde{v}} \rangle - \langle \underline{\tilde{v}}, \underline{\tilde{u}} \rangle + \langle \underline{\tilde{v}}, \underline{\tilde{v}} \rangle \\
&= \|\underline{\tilde{u}}\|^2 - \left[\langle \underline{\tilde{u}}, \underline{\tilde{v}} \rangle + \overline{\langle \underline{\tilde{u}}, \underline{\tilde{v}} \rangle}\right] + \|\underline{\tilde{v}}\|^2 \\
&= 2 - 2\text{re}\left\{\langle \underline{\tilde{u}}, \underline{\tilde{v}} \rangle\right\} \\
&= 2 - 2\left\langle \underline{\tilde{u}}, \underline{\tilde{v}} \right\rangle
\end{aligned}$$

which implies

$$\langle \underline{\tilde{u}}, \underline{\tilde{v}} \rangle \leq 1$$

or, removing the normalization,

$$\langle \underline{u}, \underline{v} \rangle \leq \|\underline{u}\| \cdot \|\underline{v}\|.$$

The same derivation holds if $\underline{u}$ is replaced by $-\underline{u}$ yielding

$$-\langle \underline{u}, \underline{v} \rangle \leq \|\underline{u}\| \cdot \|\underline{v}\|.$$

The last two equations imply

$$|\langle \underline{u}, \underline{v} \rangle| \leq \|\underline{u}\| \cdot \|\underline{v}\|.$$

In the complex case, let $\langle \underline{u}, \underline{v} \rangle = Re^{j\theta}$, and define $\underline{\tilde{v}} = \underline{v}e^{j\theta}$. Then $\langle \underline{u}, \underline{\tilde{v}} \rangle$ is real and equal to $| \langle \underline{u}, \underline{\tilde{v}} \rangle | = R > 0$. By the same derivation as above,

$$\langle \underline{u}, \underline{\tilde{v}} \rangle \leq \|\underline{u}\| \cdot \|\underline{\tilde{v}}\| = \|\underline{u}\| \cdot \|\underline{v}\|.$$

Since $\langle \underline{u}, \underline{\tilde{v}} \rangle = R = |\langle \underline{u}, \underline{\tilde{v}} \rangle| = |\langle \underline{u}, \underline{v} \rangle|$, the result is established also in the complex case.

5.9.4 Triangle Inequality

The *triangle inequality* states that the length of any side of a triangle is less than or equal to the sum of the lengths of the other two sides, with equality occurring only when the triangle degenerates to a line. In $\mathbf{C}^N$, this becomes

$$\boxed{\|\underline{u} + \underline{v}\| \leq \|\underline{u}\| + \|\underline{v}\|.}$$

We can show this quickly using the Schwarz Inequality:

$$\begin{aligned}
\|\underline{u} + \underline{v}\|^2 &= \langle \underline{u} + \underline{v}, \underline{u} + \underline{v} \rangle \\
&= \|\underline{u}\|^2 + 2\text{re}\left\{\langle \underline{u}, \underline{v} \rangle\right\} + \|\underline{v}\|^2 \\
&\leq \|\underline{u}\|^2 + 2\left|\langle \underline{u}, \underline{v} \rangle\right| + \|\underline{v}\|^2 \\
&\leq \|\underline{u}\|^2 + 2\|\underline{u}\| \cdot \|\underline{v}\| + \|\underline{v}\|^2 \\
&= \left(\|\underline{u}\| + \|\underline{v}\|\right)^2 \\
\Rightarrow \|\underline{u} + \underline{v}\| &\leq \|\underline{u}\| + \|\underline{v}\|
\end{aligned}$$

5.9.5 Triangle Difference Inequality

A useful variation on the triangle inequality is that the length of any side of a triangle is *greater* than the *absolute difference* of the lengths of the other two sides:

$$\boxed{\|\underline{u} - \underline{v}\| \geq \left|\|\underline{u}\| - \|\underline{v}\|\right|}$$

Proof: By the triangle inequality,

$$\begin{aligned}
\|\underline{v} + (\underline{u} - \underline{v})\| &\leq \|\underline{v}\| + \|\underline{u} - \underline{v}\| \\
\Rightarrow \|\underline{u} - \underline{v}\| &\geq \|\underline{u}\| - \|\underline{v}\|.
\end{aligned}$$

Interchanging $\underline{u}$ and $\underline{v}$ establishes the absolute value on the right-hand side.

5.9.6 Vector Cosine

The Cauchy-Schwarz Inequality can be written

$$\frac{|\langle \underline{u}, \underline{v}\rangle|}{\|\underline{u}\| \cdot \|\underline{v}\|} \leq 1.$$

In the case of real vectors $\underline{u}, \underline{v}$, we can always find a real number $\theta \in [0, \pi]$ which satisfies

$$\boxed{\cos(\theta) \triangleq \frac{\langle \underline{u}, \underline{v}\rangle}{\|\underline{u}\| \cdot \|\underline{v}\|}.}$$

We thus interpret θ as the *angle* between two vectors in $\mathbf{R}^N$.

5.9.7 Orthogonality

The vectors (signals) x and y[11] are said to be *orthogonal* if $\langle x, y\rangle = 0$, denoted $x \perp y$. That is to say

$$\boxed{x \perp y \Leftrightarrow \langle x, y\rangle = 0.}$$

Note that if x and y are real and orthogonal, the cosine of the angle between them is zero. In plane geometry ($N = 2$), the angle between two perpendicular lines is $\pi/2$, and $\cos(\pi/2) = 0$, as expected. More generally, orthogonality corresponds to the fact that two vectors in N-space intersect at a *right angle* and are thus *perpendicular* geometrically.

Example ($N = 2$):

Let $x = [1, 1]$ and $y = [1, -1]$, as shown in Fig. 5.8.

The inner product is $\langle x, y\rangle = 1 \cdot \overline{1} + 1 \cdot \overline{(-1)} = 0$. This shows that the vectors are *orthogonal*. As marked in the figure, the lines intersect at a right angle and are therefore perpendicular.

5.9.8 The Pythagorean Theorem in N-Space

In 2D, the Pythagorean Theorem says that when x and y are orthogonal, as in Fig. 5.8, (*i.e.*, when the vectors x and y intersect at a *right angle*), then we have

$$\|x + y\|^2 = \|x\|^2 + \|y\|^2 \qquad (x \perp y). \tag{5.1}$$

[11] Note that we have dropped the underbar notation for signals/vectors such as $\underline{x} = x$ and $\underline{y} = y$. While this is commonly done, it is now possible to confuse vectors and scalars. The context should keep everything clear. Also, symbols for scalars tend to be chosen from the beginning of the alphabet (Roman or Greek), such as $a, b, \alpha, \beta, \ldots$, while symbols for vectors/signals are normally chosen from letters near the end, such as u, v, w, x, y—all of which we have seen up to now. In later sections, the underbar notation will continue to be used when it seems to add clarity.

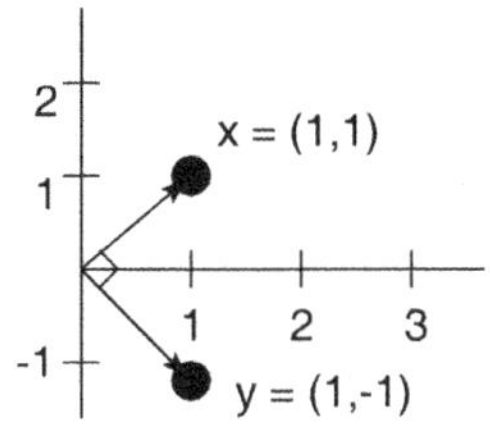

Figure 5.8: Example of two orthogonal vectors for $N = 2$.

This relationship generalizes to N dimensions, as we can easily show:

$$\begin{aligned}
\|x+y\|^2 &= \langle x+y, x+y\rangle \\
&= \langle x,x\rangle + \langle x,y\rangle + \langle y,x\rangle + \langle y,y\rangle \\
&= \|x\|^2 + \langle x,y\rangle + \overline{\langle x,y\rangle} + \|y\|^2 \\
&= \|x\|^2 + \|y\|^2 + 2\text{re}\left\{\langle x,y\rangle\right\} \qquad (5.2)
\end{aligned}$$

If $x \perp y$, then $\langle x,y\rangle = 0$ and Eq. (5.1) holds in N dimensions.

Note that the converse is not true in $\mathbf{C}^N$. That is, $\|x+y\|^2 = \|x\|^2 + \|y\|^2$ does not imply $x \perp y$ in $\mathbf{C}^N$. For a counterexample, consider $x = (j, 1)$, $y = (1, -j)$, in which case

$$\|x+y\|^2 = \|1+j, 1-j\|^2 = 4 = \|x\|^2 + \|y\|^2$$

while $\langle x,y\rangle = j \cdot 1 + 1 \cdot \overline{-j} = 2j$.

For real vectors $x, y \in \mathbf{R}^N$, the Pythagorean theorem Eq. (5.1) holds if and only if the vectors are orthogonal. To see this, note that, from Eq. (5.2), when the Pythagorean theorem holds, either x or y is zero, or $\langle x,y\rangle$ is zero or purely imaginary, by property 1 of norms (see §5.8.2 on page 79). If the inner product cannot be imaginary, it must be zero.

Note that we also have an alternate version of the Pythagorean theorem:

$$x \perp y \;\Rightarrow\; \|x-y\|^2 = \|x\|^2 + \|y\|^2$$

5.9.9 Projection

The *orthogonal projection* (or simply "projection") of $y \in \mathbf{C}^N$ onto $x \in \mathbf{C}^N$ is defined by

$$\boxed{\mathbf{P}_x(y) \stackrel{\Delta}{=} \frac{\langle y, x\rangle}{\|x\|^2}x.}$$

The complex scalar $\langle y, x\rangle / \|x\|^2$ is called the *coefficient of projection.* When projecting y onto a *unit length* vector x, the coefficient of projection is simply the inner product of y with x.

Motivation: The basic idea of orthogonal projection of y onto x is to "drop a perpendicular" from y onto x to define a new vector along x which we call the "projection" of y onto x. This is illustrated for $N = 2$ in Fig. 5.9 for $x = [4, 1]$ and $y = [2, 3]$, in which case

$$\mathbf{P}_x(y) \stackrel{\Delta}{=} \frac{\langle y, x\rangle}{\|x\|^2}x = \frac{(2 \cdot \overline{4} + 3 \cdot \overline{1})}{4^2 + 1^2}x = \frac{11}{17}x = \left[\frac{44}{17}, \frac{11}{17}\right].$$

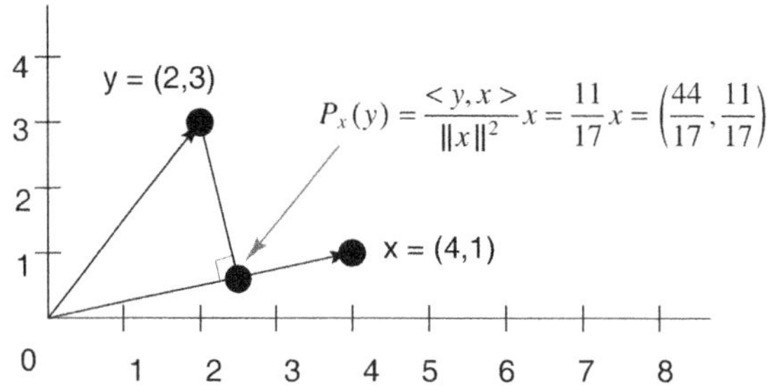

Figure 5.9: Projection of y onto x in 2D space.

Derivation: (1) Since any projection onto x must lie along the line collinear with x, write the projection as $\mathbf{P}_x(y) = \alpha x$. (2) Since by definition the *projection error* $y - \mathbf{P}_x(y)$ is orthogonal to x, we must have

$$\begin{aligned} (y - \alpha x) &\perp x \\ \Leftrightarrow \langle y - \alpha x, x\rangle &= 0 \\ \Leftrightarrow \langle y, x\rangle &= \alpha \langle x, x\rangle \\ \Leftrightarrow \alpha &= \frac{\langle y, x\rangle}{\langle x, x\rangle} = \frac{\langle y, x\rangle}{\|x\|^2}. \end{aligned}$$

Thus,

$$\mathbf{P}_x(y) = \frac{\langle y, x\rangle}{\|x\|^2}x.$$

See §I.3.3 on page 277 for illustration of orthogonal projection in matlab.

5.10 Signal Reconstruction from Projections

We now know how to project a signal onto other signals. We now need to learn how to reconstruct a signal $x \in \mathbf{C}^N$ from its projections onto N different vectors $\underline{s}_k \in \mathbf{C}^N$, $k = 0, 1, 2, \ldots, N-1$. This will give us the *inverse DFT* operation (or the inverse of whatever transform we are working with).

As a simple example, consider the projection of a signal $x \in \mathbf{C}^N$ onto the rectilinear *coordinate axes* of $\mathbf{C}^N$. The coordinates of the projection onto the 0th coordinate axis are simply $(x_0, 0, \ldots, 0)$. The projection along coordinate axis 1 has coordinates $(0, x_1, 0, \ldots, 0)$, and so on. The original signal x is then clearly the *vector sum* of its projections onto the coordinate axes:

$$\begin{aligned} x &= (x_0, \ldots, x_{N-1}) \\ &= (x_0, 0, \ldots, 0) + (0, x_1, 0, \ldots, 0) + \cdots (0, \ldots, 0, x_{N-1}) \end{aligned}$$

To make sure the previous paragraph is understood, let's look at the details for the case $N = 2$. We want to project an arbitrary two-sample signal $x = (x_0, x_1)$ onto the coordinate axes in 2D. A coordinate axis can be generated by multiplying any nonzero vector by scalars. The horizontal axis can be represented by any vector of the form $\underline{e}_0 = (\alpha, 0)$, $\alpha \neq 0$ while the vertical axis can be represented by any vector of the form $\underline{e}_1 = (0, \beta)$, $\beta \neq 0$. For maximum simplicity, let's choose

$$\begin{aligned} \underline{e}_0 &\triangleq [1, 0], \\ \underline{e}_1 &\triangleq [0, 1]. \end{aligned}$$

The projection of x onto $\underline{e}_0$ is, by definition,

$$\begin{aligned} \mathbf{P}_{\underline{e}_0}(x) &\triangleq \frac{\langle x, \underline{e}_0 \rangle}{\|\underline{e}_0\|^2} \underline{e}_0 \\ &= \langle x, \underline{e}_0 \rangle \, \underline{e}_0 = \langle [x_0, x_1], [1, 0] \rangle \, \underline{e}_0 = (x_0 \cdot \overline{1} + x_1 \cdot \overline{0}) \underline{e}_0 = x_0 \underline{e}_0 \\ &= [x_0, 0]. \end{aligned}$$

Similarly, the projection of x onto $\underline{e}_1$ is

$$\begin{aligned}
\mathbf{P}_{\underline{e}_1}(x) &\triangleq \frac{\langle x, \underline{e}_1 \rangle}{\|\underline{e}_1\|^2}\underline{e}_1 \\
&= \langle x, \underline{e}_1 \rangle\, \underline{e}_1 = \langle [x_0, x_1], [0, 1] \rangle\, \underline{e}_1 = (x_0 \cdot \overline{0} + x_1 \cdot \overline{1})\underline{e}_1 = x_1 \underline{e}_1 \\
&= [0, x_1].
\end{aligned}$$

The *reconstruction* of x from its projections onto the coordinate axes is then the *vector sum of the projections*:

$$x = \mathbf{P}_{\underline{e}_0}(x) + \mathbf{P}_{\underline{e}_1}(x) = x_0 \underline{e}_0 + x_1 \underline{e}_1 \triangleq x_0 \cdot [1, 0] + x_1 \cdot [0, 1] = (x_0, x_1)$$

The projection of a vector onto its coordinate axes is in some sense trivial because the very meaning of the *coordinates* is that they are scalars x_n to be applied to the *coordinate vectors* $\underline{e}_n$ in order to form an arbitrary vector $x \in \mathbf{C}^N$ as a *linear combination* of the coordinate vectors:

$$x \triangleq x_0 \underline{e}_0 + x_1 \underline{e}_1 + \cdots + x_{N-1} \underline{e}_{N-1}$$

Note that the coordinate vectors are *orthogonal*. Since they are also unit length, $\|\underline{e}_n\| = 1$, we say that the coordinate vectors $\{\underline{e}_n\}_{n=0}^{N-1}$ are *orthonormal*.

5.10.1 Changing Coordinates

What's more interesting is when we project a signal x onto a set of vectors *other than* the coordinate set. This can be viewed as a *change of coordinates* in $\mathbf{C}^N$. In the case of the DFT, the new vectors will be chosen to be *sampled complex sinusoids.*

An Example of Changing Coordinates in 2D

As a simple example, let's pick the following pair of new coordinate vectors in 2D:

$$\begin{aligned}
\underline{s}_0 &\triangleq [1, 1] \\
\underline{s}_1 &\triangleq [1, -1]
\end{aligned}$$

These happen to be the DFT sinusoids for $N = 2$ having frequencies $f_0 = 0$ ("dc") and $f_1 = f_s/2$ (half the sampling rate). (The sampled complex sinusoids of the DFT reduce to real numbers only for $N = 1$ and $N = 2$.) We already showed in an earlier example that these vectors are *orthogonal.* However, they are not orthonormal since the norm is $\sqrt{2}$ in each case.

Let's try projecting x onto these vectors and seeing if we can reconstruct by summing the projections.

The projection of x onto $\underline{s}_0$ is, by definition,[12]

$$\begin{aligned}\mathbf{P}_{\underline{s}_0}(x) &\isdef \frac{\langle x,\underline{s}_0\rangle}{\|\underline{s}_0\|^2}\underline{s}_0 = \frac{\langle [x_0,x_1],[1,1]\rangle}{2}\underline{s}_0 \\ &= \frac{(x_0\cdot\overline{1}+x_1\cdot\overline{1})}{2}\underline{s}_0 = \frac{x_0+x_1}{2}\underline{s}_0.\end{aligned}$$

Similarly, the projection of x onto $\underline{s}_1$ is

$$\begin{aligned}\mathbf{P}_{\underline{s}_1}(x) &\isdef \frac{\langle x,\underline{s}_1\rangle}{\|\underline{s}_1\|^2}\underline{s}_1 = \frac{\langle [x_0,x_1],[1,-1]\rangle}{2}\underline{s}_1 \\ &= \frac{(x_0\cdot\overline{1}-x_1\cdot\overline{1})}{2}\underline{s}_1 = \frac{x_0-x_1}{2}\underline{s}_1.\end{aligned}$$

The sum of these projections is then

$$\begin{aligned}\mathbf{P}_{\underline{s}_0}(x)+\mathbf{P}_{\underline{s}_1}(x) &= \frac{x_0+x_1}{2}\underline{s}_0+\frac{x_0-x_1}{2}\underline{s}_1 \\ &\isdef \frac{x_0+x_1}{2}(1,1)+\frac{x_0-x_1}{2}(1,-1) \\ &= \left(\frac{x_0+x_1}{2}+\frac{x_0-x_1}{2}, \frac{x_0+x_1}{2}-\frac{x_0-x_1}{2}\right) \\ &= (x_0,x_1) \isdef x.\end{aligned}$$

It worked!

Projection onto Linearly Dependent Vectors

Now consider another example:

$$\begin{aligned}\underline{s}_0 &\isdef [1,1], \\ \underline{s}_1 &\isdef [-1,-1].\end{aligned}$$

The projections of $x=[x_0,x_1]$ onto these vectors are

$$\begin{aligned}\mathbf{P}_{\underline{s}_0}(x) &= \frac{x_0+x_1}{2}\underline{s}_0, \\ \mathbf{P}_{\underline{s}_1}(x) &= -\frac{x_0+x_1}{2}\underline{s}_1.\end{aligned}$$

[12]Note that, in this section, x denotes an entire *signal* while $x_n = x(n)$ denotes the nth *sample* of that signal. It would be clearer to use $x(n)$, but the expressions below would become messier. In other contexts, outside of this section, x_i might instead denote the ith *signal* $\underline{x}_i$ from a set of signals.

The sum of the projections is

$$\begin{aligned}
\mathbf{P}_{\underline{s}_0}(x) + \mathbf{P}_{\underline{s}_1}(x) &= \frac{x_0 + x_1}{2}\underline{s}_0 - \frac{x_0 + x_1}{2}\underline{s}_1 \\
&\triangleq \frac{x_0 + x_1}{2}(1,1) - \frac{x_0 + x_1}{2}(-1,-1) \\
&= (x_0 + x_1, x_0 + x_1) \neq x.
\end{aligned}$$

Something went wrong, but what? It turns out that a set of N vectors can be used to reconstruct an arbitrary vector in $\mathbf{C}^N$ from its projections only if they are *linearly independent*. In general, a set of vectors is linearly independent if none of them can be expressed as a linear combination of the others in the set. What this means intuitively is that they must "point in different directions" in N-space. In this example $s_1 = -s_0$ so that they lie along the *same line* in 2-space. As a result, they are linearly *dependent*: one is a linear combination of the other ($s_1 = (-1)s_0$).

Projection onto Non-Orthogonal Vectors

Consider this example:

$$\begin{aligned}
\underline{s}_0 &\triangleq [1,1] \\
\underline{s}_1 &\triangleq [0,1]
\end{aligned}$$

These point in different directions, but they are not orthogonal. What happens now? The projections are

$$\begin{aligned}
\mathbf{P}_{\underline{s}_0}(x) &= \frac{x_0 + x_1}{2}\underline{s}_0 \\
\mathbf{P}_{\underline{s}_1}(x) &= x_1 \underline{s}_1.
\end{aligned}$$

The sum of the projections is

$$\begin{aligned}
\mathbf{P}_{\underline{s}_0}(x) + \mathbf{P}_{\underline{s}_1}(x) &= \frac{x_0 + x_1}{2}\underline{s}_0 + x_1\underline{s}_1 \\
&\triangleq \frac{x_0 + x_1}{2}(1,1) + x_1 \cdot (0,1) \\
&= \left(\frac{x_0 + x_1}{2}, \frac{x_0 + 3x_1}{2}\right) \\
&\neq x.
\end{aligned}$$

So, even though the vectors are linearly independent, the sum of projections onto them does not reconstruct the original vector. Since the sum of projections worked in the orthogonal case, and since orthogonality implies linear independence, we might conjecture at this point that the sum of projections onto a set of N vectors will reconstruct the original vector only when the vector set is *orthogonal*, and this is true, as we will show.

It turns out that one can apply an orthogonalizing process, called *Gram-Schmidt orthogonalization* to any N linearly independent vectors in $\mathbf{C}^N$ so as to form an orthogonal set which will always work. This will be derived in Section 5.10.4.

Obviously, there must be at least N vectors in the set. Otherwise, there would be too few *degrees of freedom* to represent an arbitrary $x \in \mathbf{C}^N$. That is, given the N coordinates $\{u(n)\}_{n=0}^{N-1}$ of x (which are scale factors relative to the coordinate vectors $\underline{e}_n$ in $\mathbf{C}^N$), we have to find at least N coefficients of projection (which we may think of as coordinates relative to new coordinate vectors $\underline{s}_k$). If we compute only $M < N$ coefficients, then we would be mapping a set of N complex numbers to $M < N$ numbers. Such a mapping cannot be invertible in general. It also turns out N linearly independent vectors is always sufficient. The next section will summarize the general results along these lines.

5.10.2 General Conditions

This section summarizes and extends the above derivations in a more formal manner (following portions of chapter 4 of [45]). In particular, we establish that the sum of projections of $x \in \mathbf{C}^N$ onto N vectors $\underline{s}_k \in \mathbf{C}^N$ will give back the original vector x whenever the set $\{\underline{s}_k\}_0^{N-1}$ is an *orthogonal basis* for $\mathbf{C}^N$.

Definition: A set of vectors is said to form a *vector space* if, given any two members x and y from the set, the vectors $x + y$ and cx are also in the set, where $c \in \mathbf{C}^N$ is any scalar.

Definition: The set of all N-dimensional complex vectors is denoted $\mathbf{C}^N$. That is, $\mathbf{C}^N$ consists of all vectors $\underline{x} = (x_0, \ldots, x_{N-1})$ defined as a list of N complex numbers $x_i \in \mathbf{C}$.

Theorem: $\mathbf{C}^N$ is a *vector space* under elementwise addition and multiplication by complex scalars.

Proof: This is a special case of the following more general theorem.

Theorem: Let $M \leq N$ be an integer greater than 0. Then the set of all linear combinations of M vectors from $\mathbf{C}^N$ forms a vector space under elementwise addition and multiplication by complex scalars.

Proof: Let the original set of M vectors be denoted $\underline{s}_0, \ldots, \underline{s}_{M-1}$. Form

$$x_0 = \alpha_0 \underline{s}_0 + \cdots + \alpha_{M-1} \underline{s}_{M-1}$$

as a particular linear combination of $\{\underline{s}_i\}_{i=0}^{M-1}$. Then

$$cx_0 = c\alpha_0\underline{s}_0 + \cdots + c\alpha_{M-1}\underline{s}_{M-1}$$

is also a linear combination of $\{\underline{s}_i\}_{i=0}^{M-1}$, since complex numbers are closed under multiplication ($c\alpha_i \in \mathbf{C}$ for each i). Now, given any second linear combination of $\{\underline{s}_i\}_{i=0}^{M-1}$,

$$x_1 = \beta_0\underline{s}_0 + \cdots + \beta_{M-1}\underline{s}_{M-1},$$

the sum is

$$\begin{aligned} x_0 + x_1 &= (\alpha_0\underline{s}_0 + \cdots \alpha_{M-1}\underline{s}_{M-1}) + (\beta_0\underline{s}_0 + \cdots \beta_{M-1}\underline{s}_{M-1}) \\ &= (\alpha_0 + \beta_0)\underline{s}_0 + \cdots + (\alpha_{M-1} + \beta_{M-1})\underline{s}_{M-1} \end{aligned}$$

which is yet another linear combination of the original vectors (since complex numbers are closed under addition). Since we have shown that scalar multiples and vector sums of linear combinations of the original M vectors from $\mathbf{C}^N$ are also linear combinations of those same original M vectors from $\mathbf{C}^N$, we have that the defining properties of a vector space are satisfied. □

Note that the closure of vector addition and scalar multiplication are "inherited" from the closure of complex numbers under addition and multiplication.

Corollary: The set of all linear combinations of M *real* vectors $x \in \mathbf{R}^N$, using real scalars $\alpha_i \in \mathbf{R}$, form a vector space.

Definition: The set of all linear combinations of a set of M *complex* vectors from $\mathbf{C}^N$, using complex scalars, is called a *complex vector space* of dimension M.

Definition: The set of all linear combinations of a set of M *real* vectors from $\mathbf{R}^N$, using real scalars, is called a *real vector space* of dimension M.

Definition: If a vector space consists of the set of all linear combinations of a finite set of vectors $\underline{s}_0, \ldots, \underline{s}_{M-1}$, then those vectors are said to *span* the space.

Example: The *coordinate vectors* in $\mathbf{C}^N$ span $\mathbf{C}^N$ since every vector $x \in \mathbf{C}^N$ can be expressed as a linear combination of the coordinate vectors as

$$x = x_0\underline{e}_0 + x_1\underline{e}_1 + \cdots + x_{N-1}\underline{e}_{N-1}$$

where $x_i \in \mathbf{C}$, and

$$\begin{aligned} \underline{e}_0 &= (1,0,0,\ldots,0), \\ \underline{e}_1 &= (0,1,0,\ldots,0), \\ \underline{e}_2 &= (0,0,1,0,\ldots,0), \text{ and so on up to} \\ \underline{e}_{N-1} &= (0,\ldots,0,1). \end{aligned}$$

Definition: The vector space spanned by a set of $M < N$ vectors from $\mathbf{C}^N$ is called an M-dimensional *subspace* of $\mathbf{C}^N$.

Definition: A vector $\underline{s} \in \mathbf{C}^N$ is said to be *linearly dependent* on a set of $M \leq N$ vectors $\underline{s}_i \in \mathbf{C}^N$, $m = 0,\ldots,M-1$, if $\underline{s}$ can be expressed as a linear combination of those M vectors.

Thus, $\underline{s}$ is linearly dependent on $\{\underline{s}_i\}_{i=0}^{M-1}$ if there exist scalars $\{\alpha_i\}_{i=0}^{M-1}$ such that $\underline{s} = \alpha_0\underline{s}_0 + \alpha_1\underline{s}_1 + \cdots + \alpha_{M-1}\underline{s}_{M-1}$. Note that the zero vector is linearly dependent on every collection of vectors.

Theorem: (i) If $\underline{s}_0,\ldots,\underline{s}_{M-1}$ span a vector space, and if one of them, say $\underline{s}_m$, is linearly dependent on the others, then the same vector space is spanned by the set obtained by omitting $\underline{s}_m$ from the original set. (ii) If $\underline{s}_0,\ldots,\underline{s}_{M-1}$ span a vector space, we can always select from these a linearly independent set that spans the same space.

Proof: Any x in the space can be represented as a linear combination of the vectors $\underline{s}_0,\ldots,\underline{s}_{M-1}$. By expressing $\underline{s}_m$ as a linear combination of the other vectors in the set, the linear combination for x becomes a linear combination of vectors other than $\underline{s}_m$. Thus, $\underline{s}_m$ can be eliminated from the set, proving (i). To prove (ii), we can define a procedure for forming the required subset of the original vectors: First, assign $\underline{s}_0$ to the set. Next, check to see if $\underline{s}_0$ and $\underline{s}_1$ are linearly dependent. If so (*i.e.*, $\underline{s}_1$ is a scalar times $\underline{s}_0$), then discard $\underline{s}_1$; otherwise assign it also to the new set. Next, check to see if $\underline{s}_2$ is linearly dependent on the vectors in the new set. If it is (*i.e.*, $\underline{s}_2$ is some linear combination of $\underline{s}_0$ and $\underline{s}_1$) then discard it; otherwise assign it also to the new set. When this procedure terminates after processing $\underline{s}_{M-1}$, the new set will contain only linearly independent vectors which span the original space.

Definition: A set of linearly independent vectors which spans a vector space is called a *basis* for that vector space.

Definition: The set of coordinate vectors in $\mathbf{C}^N$ is called the *natural basis* for $\mathbf{C}^N$, where the nth basis vector is

$$\underline{e}_n = [\,0\ \cdots\ 0\ \underbrace{1}_{n\text{th}}\ 0\ \cdots\ 0].$$

Theorem: The linear combination expressing a vector in terms of basis vectors for a vector space is *unique*.

Proof: Suppose a vector $x \in \mathbf{C}^N$ can be expressed in two different ways as a linear combination of basis vectors $\underline{s}_0, \ldots, \underline{s}_{N-1}$:

$$\begin{aligned} x &= \alpha_0 \underline{s}_0 + \cdots \alpha_{N-1} \underline{s}_{N-1} \\ &= \beta_0 \underline{s}_0 + \cdots \beta_{N-1} \underline{s}_{N-1} \end{aligned}$$

where $\alpha_i \neq \beta_i$ for at least one value of $i \in [0, N-1]$. Subtracting the two representations gives

$$\underline{0} = (\alpha_0 - \beta_0)\underline{s}_0 + \cdots + (\alpha_{N-1} - \beta_{N-1})\underline{s}_{N-1}.$$

Since the vectors are linearly independent, it is not possible to cancel the nonzero vector $(\alpha_i - \beta_i)\underline{s}_i$ using some linear combination of the other vectors in the sum. Hence, $\alpha_i = \beta_i$ for all $i = 0, 1, 2, \ldots, N-1$.

Note that while the linear combination relative to a particular basis is unique, the choice of basis vectors is not. For example, given any basis set in $\mathbf{C}^N$, a new basis can be formed by *rotating* all vectors in $\mathbf{C}^N$ by the same angle. In this way, an infinite number of basis sets can be generated.

As we will soon show, the DFT can be viewed as a *change of coordinates* from coordinates relative to the *natural basis* in $\mathbf{C}^N$, $\{\underline{e}_n\}_{n=0}^{N-1}$, to coordinates relative to the *sinusoidal basis* for $\mathbf{C}^N$, $\{\underline{s}_k\}_{k=0}^{N-1}$, where $\underline{s}_k(n) = e^{j\omega_k t_n}$. The sinusoidal basis set for $\mathbf{C}^N$ consists of length N sampled complex sinusoids at frequencies $\omega_k = 2\pi k f_s/N, k = 0, 1, 2, \ldots, N-1$. Any scaling of these vectors in $\mathbf{C}^N$ by complex scale factors could also be chosen as the sinusoidal basis (*i.e.*, any nonzero amplitude and any phase will do). However, for simplicity, we will only use unit-amplitude, zero-phase complex sinusoids as the Fourier "frequency-domain" basis set. To summarize this paragraph, the time-domain samples of a signal are its coordinates relative to the natural basis for $\mathbf{C}^N$, while its spectral coefficients are the coordinates of the signal relative to the sinusoidal basis for $\mathbf{C}^N$.

Theorem: Any two bases of a vector space contain the same number of vectors.

Proof: Left as an exercise (or see [45]).

Definition: The number of vectors in a basis for a particular space is called the *dimension* of the space. If the dimension is N, the space is said to be an N *dimensional space*, or N*-space*.

In this book, we will only consider finite-dimensional vector spaces in any detail. However, the discrete-time Fourier transform (DTFT) and Fourier transform (FT) both require infinite-dimensional basis sets, because there is an infinite number of points in both the time and frequency domains. (See Appendix B for details regarding the FT and DTFT.)

5.10.3 Signal/Vector Reconstruction from Projections

We now arrive finally at the main desired result for this section:

Theorem: The projections of any vector $x \in \mathbf{C}^N$ onto any orthogonal basis set for $\mathbf{C}^N$ can be summed to reconstruct x exactly.

Proof: Let $\{\underline{s}_0, \ldots, \underline{s}_{N-1}\}$ denote any orthogonal basis set for $\mathbf{C}^N$. Then since x is in the space spanned by these vectors, we have

$$x = \alpha_0 \underline{s}_0 + \alpha_1 \underline{s}_1 + \cdots + \alpha_{N-1} \underline{s}_{N-1} \tag{5.3}$$

for some (unique) scalars $\alpha_0, \ldots, \alpha_{N-1}$. The projection of x onto $\underline{s}_k$ is equal to

$$\mathbf{P}_{\underline{s}_k}(x) = \alpha_0 \mathbf{P}_{\underline{s}_k}(\underline{s}_0) + \alpha_1 \mathbf{P}_{\underline{s}_k}(\underline{s}_1) + \cdots + \alpha_{N-1} \mathbf{P}_{\underline{s}_k}(\underline{s}_{N-1})$$

(using the linearity of the projection operator which follows from linearity of the inner product in its first argument). Since the basis vectors are orthogonal, the projection of $\underline{s}_l$ onto $\underline{s}_k$ is zero for $l \neq k$:

$$\mathbf{P}_{\underline{s}_k}(\underline{s}_l) \triangleq \frac{\langle \underline{s}_l, \underline{s}_k \rangle}{\| \underline{s}_k \|^2} \underline{s}_k = \begin{cases} \underline{0}, & l \neq k \\ \underline{s}_k, & l = k. \end{cases}$$

We therefore obtain

$$\mathbf{P}_{\underline{s}_k}(x) = 0 + \cdots + 0 + \alpha_k \mathbf{P}_{\underline{s}_k}(\underline{s}_k) + 0 + \cdots + 0 = \alpha_k \underline{s}_k.$$

Therefore, the sum of projections onto the vectors $\underline{s}_k$, $k = 0, 1, \ldots, N-1$, is just the linear combination of the $\underline{s}_k$ which forms x:

$$\sum_{k=0}^{N-1} \mathbf{P}_{\underline{s}_k}(x) = \sum_{k=0}^{N-1} \alpha_k \underline{s}_k = x$$

by Eq. (5.3). □

5.10.4 Gram-Schmidt Orthogonalization

Recall from the end of §5.10 above that an *orthonormal* set of vectors is a set of *unit-length* vectors that are mutually *orthogonal*. In other words, orthonormal vector set is just an orthogonal vector set in which each vector $\underline{s}_i$ has been normalized to unit length $\underline{s}_i / \| \underline{s}_i \|$.

Theorem: Given a set of N linearly independent vectors $\underline{s}_0, \ldots, \underline{s}_{N-1}$ from $\mathbf{C}^N$, we can construct an *orthonormal* set $\underline{\tilde{s}}_0, \ldots, \underline{\tilde{s}}_{N-1}$ which are linear combinations of the original set and which span the same space.

Proof: We prove the theorem by constructing the desired orthonormal set $\{\underline{\tilde{s}}_k\}$ sequentially from the original set $\{\underline{s}_k\}$. This procedure is known as *Gram-Schmidt orthogonalization.*

First, note that $\underline{s}_k \neq \underline{0}$ for all k, since $\underline{0}$ is linearly dependent on every vector. Therefore, $\| \underline{s}_k \| \neq 0$.

- Set $\underline{\tilde{s}}_0 \stackrel{\Delta}{=} \frac{\underline{s}_0}{\| \underline{s}_0 \|}$.
- Define $\underline{x}_1$ as $\underline{s}_1$ minus the projection of $\underline{s}_1$ onto $\underline{\tilde{s}}_0$:

$$\underline{x}_1 \stackrel{\Delta}{=} \underline{s}_1 - \mathbf{P}_{\underline{\tilde{s}}_0}(\underline{s}_1) = \underline{s}_1 - \langle \underline{s}_1, \underline{\tilde{s}}_0 \rangle \, \underline{\tilde{s}}_0$$

  The vector $\underline{x}_1$ is orthogonal to $\underline{\tilde{s}}_0$ by construction. (We subtracted out the part of $\underline{s}_1$ that wasn't orthogonal to $\underline{\tilde{s}}_0$.) Also, since $\underline{s}_1$ and $\underline{s}_0$ are linearly independent, we have $\| \underline{x}_1 \| \neq 0$.

- Set $\underline{\tilde{s}}_1 \stackrel{\Delta}{=} \frac{\underline{x}_1}{\| \underline{x}_1 \|}$ (*i.e.*, normalize the result of the preceding step).
- Define $\underline{x}_2$ as $\underline{s}_2$ minus the projection of $\underline{s}_2$ onto $\underline{\tilde{s}}_0$ and $\underline{\tilde{s}}_1$:

$$\underline{x}_2 \stackrel{\Delta}{=} \underline{s}_2 - \mathbf{P}_{\underline{\tilde{s}}_0}(\underline{s}_2) - \mathbf{P}_{\underline{\tilde{s}}_1}(\underline{s}_2) = \underline{s}_2 - \langle \underline{s}_2, \underline{\tilde{s}}_0 \rangle \, \underline{\tilde{s}}_0 - \langle \underline{s}_2, \underline{\tilde{s}}_1 \rangle \, \underline{\tilde{s}}_1$$

- Normalize: $\underline{\tilde{s}}_2 \stackrel{\Delta}{=} \frac{\underline{x}_2}{\| \underline{x}_2 \|}$.
- Continue this process until $\underline{\tilde{s}}_{N-1}$ has been defined.

The Gram-Schmidt orthogonalization procedure will construct an orthonormal basis from any set of N linearly independent vectors. Obviously, by skipping the normalization step, we could also form simply an orthogonal basis. The key ingredient of this procedure is that each new basis vector is obtained by subtracting out the projection of the next linearly independent vector onto the vectors accepted so far into the set. We may say that each new linearly independent vector $\underline{s}_k$ is projected onto the *subspace* spanned by the vectors $\{\underline{\tilde{s}}_0, \ldots, \underline{\tilde{s}}_{k-1}\}$, and any nonzero projection in that subspace is subtracted out of $\underline{s}_k$ to make the new vector orthogonal to the entire subspace. In other words, we retain only that portion of each new vector $\underline{s}_k$ which "points along" a new dimension. The first direction is arbitrary and is determined by whatever vector we choose first ($\underline{s}_0$ here). The next vector is forced to be orthogonal to the first. The second is forced to be orthogonal to the first two (and thus to the 2D subspace spanned by them), and so on.

This chapter can be considered an introduction to some important concepts of *linear algebra*. The student is invited to pursue further reading in any textbook on linear algebra, such as [45].[13]

Matlab/Octave examples related to this chapter appear in Appendix I.

5.11 Signal Projection Problems

1. Find the components (sample values) of the sinusoids $x_k(n) \triangleq e^{j\omega_k nT}$, where $\omega_k T = 2\pi k/N$, and $n = 0, 1, \ldots, N-1$, $k = 0, 1, \ldots, N-1$, and the dimension of the signal space is $N = 4$. *Example:* $x_0 = [1, 1, 1, 1]$.

2. For each k, and with $N = 4$, plot (on the unit circle in the complex plane) the *generators*

$$W_N^k \triangleq e^{j2\pi\frac{k}{N}}, \quad k = 0, 1, \ldots, N-1,$$

of the sampled sinusoids $x_k(n) = \left[W_N^k\right]^n$ defined in the previous problem. From looking at this diagram, find

$$\sum_{k=0}^{N-1} W_N^k.$$

3. Verify the *orthogonality* of the set of signal vectors $\{x_k\}$ defined in the previous problem by computing the following *inner products*: $\langle x_0, x_1\rangle$, $\langle x_0, x_2\rangle$, $\langle x_0, x_3\rangle$, $\langle x_1, x_2\rangle$, $\langle x_1, x_3\rangle$, $\langle x_2, x_3\rangle$. Why do we need no more combinations than these six?

4. Show that each $x_k(n)$ (again as defined in the previous problems) has *norm*, or "length", equal to

$$\|x_k\| = \sqrt{\langle x_k, x_k\rangle} = 2.$$

5. Find the orthogonal projection of the signal $y(n) = n$ (i.e., $y = [0, 1, 2, 3]$), onto each of the signals x_k, $k = 0, 1, 2, 3$, from the previous problem.
Example:

$$y_{x_0}(n) \triangleq \frac{\langle y, x_0\rangle}{\|x_0\|^2} x_0(n) = \frac{6}{4} x_0(n) = [1.5, 1.5, 1.5, 1.5]$$

[13] An excellent collection of free downloadable course videos by Prof. Strang at MIT is available at `http://web.mit.edu/18.06/www/Video/video-fall-99-new.html`

6. Add the orthogonal projections together from the previous problem. That is, find the sum

$$s(n) \stackrel{\Delta}{=} y_{x_0}(n) + y_{x_1}(n) + y_{x_2}(n) + y_{x_3}(n)$$

Explain the significance of the above operations in the spectrum analysis context. What are we doing from a geometric perspective?

7. Find the Discrete Fourier Transform (DFT) of the sequence $y = [0, 1, 2, 3]$.

8. Show

$$\sum_{n=0}^{N-1} e^{j2\pi \frac{k}{N} n} = \begin{cases} N, & k = 0 \\ 0, & k \neq 0 \end{cases}$$

That is, show that every complex sinusoid with nonzero frequency $\omega_k T = 2\pi k/N$ is zero mean. [Hint: Consider a vector plot in the z plane.]

9. Define

$$\begin{aligned} c_k(n) &= \cos(\omega_k n) \\ s_k(n) &= \sin(\omega_k n) \\ e_k(n) &= e^{j\omega_k n} \end{aligned}$$

where $\omega_k T \stackrel{\Delta}{=} 2\pi k/N$, and integers n and k range from 0 to $N - 1$. Show that, given $e_k \perp e_l$, for all $k \neq l$, we have

$$\begin{aligned} c_k &\perp c_l, \quad k \neq l, \quad k \neq N - l \\ s_k &\perp s_l, \quad k \neq l, \quad k \neq N - l \\ s_k &\perp c_l, \quad k \neq l, \quad k \neq N - l. \end{aligned}$$

[Hint: Use Euler's identity.]

10. Show that the restrictions $k \neq l$, and $k \neq N - l$ in the previous problem can be removed in the case of $s_k \perp c_l$.

Chapter 6

Derivation of the Discrete Fourier Transform (DFT)

This chapter derives the Discrete Fourier Transform (DFT) as a projection of a length N signal x onto the set of N sampled complex sinusoids generated by the Nth roots of unity.

6.1 Geometric Series

Recall that for any complex number $z_1 \in \mathbf{C}$, the signal

$$x(n) \stackrel{\Delta}{=} z_1^n, \quad n = 0, 1, 2, \ldots,$$

defines a *geometric sequence*, *i.e.*, each term is obtained by multiplying the previous term by the (complex) constant z_1. A *geometric series* is the *sum* of a geometric sequence:

$$S_N(z_1) \stackrel{\Delta}{=} \sum_{n=0}^{N-1} z_1^n = 1 + z_1 + z_1^2 + z_1^3 + \cdots + z_1^{N-1}$$

If $z_1 \neq 1$, the sum can be expressed in *closed form*:

$$\boxed{S_N(z_1) = \frac{1 - z_1^N}{1 - z_1}} \qquad (z_1 \neq 1)$$

Proof: We have

$$\begin{aligned}
S_N(z_1) &\stackrel{\Delta}{=} 1 + z_1 + z_1^2 + z_1^3 + \cdots + z_1^{N-1} \\
z_1 S_N(z_1) &= \quad\;\; z_1 + z_1^2 + z_1^3 + \cdots + z_1^{N-1} + z_1^N \\
\Rightarrow\ S_N(z_1) - z_1 S_N(z_1) &= 1 - z_1^N \\
\Rightarrow\ S_N(z_1) &= \frac{1 - z_1^N}{1 - z_1}.
\end{aligned}$$

When $z_1 = 1$, $S_N(1) = N$, by inspection of the definition of $S_N(z_1)$.

6.2 Orthogonality of Sinusoids

A key property of sinusoids is that they are *orthogonal at different frequencies*. That is,

$$\omega_1 \neq \omega_2 \implies A_1 \sin(\omega_1 t + \phi_1) \perp A_2 \sin(\omega_2 t + \phi_2).$$

This is true whether they are complex or real, and whatever amplitude and phase they may have. All that matters is that the frequencies be different. Note, however, that the durations must be infinity (in general).

For length N *sampled* sinusoidal signal segments, such as used by the DFT, exact orthogonality holds only for the *harmonics of the sampling-rate-divided-by-N*, *i.e.*, only for the frequencies (in Hz)

$$f_k = k\frac{f_s}{N}, \quad k = 0, 1, 2, 3, \ldots, N-1.$$

These are the only frequencies that have a *whole number of periods in N samples* (depicted in Fig. 6.2 for $N = 8$).[1]

The complex sinusoids corresponding to the frequencies f_k are

$$s_k(n) \stackrel{\Delta}{=} e^{j\omega_k nT}, \quad \omega_k \stackrel{\Delta}{=} k\frac{2\pi}{N} f_s, \quad k = 0, 1, 2, \ldots, N-1.$$

These sinusoids are generated by the Nth *roots of unity* in the complex plane.

6.2.1 Nth Roots of Unity

As introduced in §3.12 on page 28, the complex numbers

$$W_N^k \stackrel{\Delta}{=} e^{j\omega_k T} \stackrel{\Delta}{=} e^{jk2\pi(f_s/N)T} = e^{jk2\pi/N}, \quad k = 0, 1, 2, \ldots, N-1,$$

[1]The Matlab code for generating this figure is given in §I.4.1.

are called the *Nth roots of unity* because each of them satisfies

$$\left[W_N^k\right]^N = \left[e^{j\omega_k T}\right]^N = \left[e^{jk2\pi/N}\right]^N = e^{jk2\pi} = 1.$$

In particular, W_N is called a *primitive Nth root of unity.*[2]

The Nth roots of unity are plotted in the complex plane in Fig. 6.1 for $N = 8$. It is easy to find them graphically by dividing the unit circle into N equal parts using N points, with one point anchored at $z = 1$, as indicated in Fig. 6.1. When N is even, there will be a point at $z = -1$ (corresponding to a sinusoid with frequency at exactly half the sampling rate), while if N is odd, there is no point at $z = -1$.

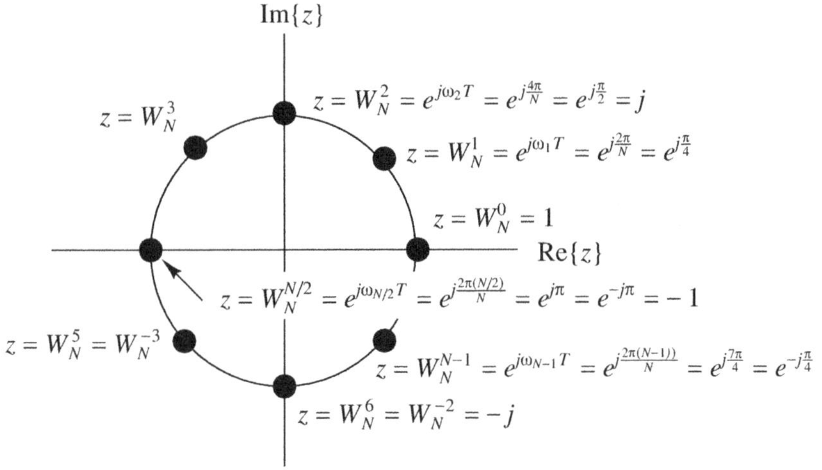

Figure 6.1: The N roots of unity for $N = 8$.

6.2.2 DFT Sinusoids

The sampled sinusoids generated by integer powers of the N roots of unity are plotted in Fig. 6.2. These are the sampled sinusoids $(W_N^k)^n = e^{j2\pi kn/N} = e^{j\omega_k nT}$ used by the DFT. Note that taking successively higher integer powers of the point W_N^k on the unit circle *generates* samples of the kth DFT sinusoid, giving $[W_N^k]^n$, $n = 0, 1, 2, \ldots, N-1$. The kth sinusoid generator W_N^k is in turn the kth Nth root of unity (kth power of the primitive Nth root of unity W_N).

[2]The notations W_N, W_N^k, and W_N^{nk} are common in the digital signal processing literature. Sometimes W_N is defined with a negative exponent, *i.e.*, $W_N \triangleq \exp(-j2\pi/N)$.

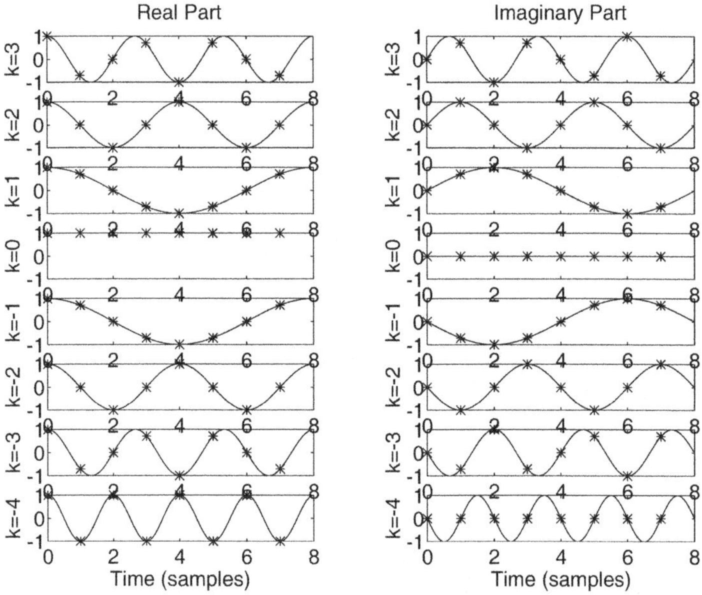

Figure 6.2: Complex sinusoids used by the DFT for $N = 8$.

Note that in Fig. 6.2 the range of k is taken to be $[-N/2, N/2-1] = [-4, 3]$ instead of $[0, N-1] = [0, 7]$. This is the most "physical" choice since it corresponds with our notion of "negative frequencies." However, we may add any integer multiple of N to k without changing the sinusoid indexed by k. In other words, $k \pm mN$ refers to the same sinusoid $\exp(j\omega_k nT)$ for all integers m.

6.3 Orthogonality of the DFT Sinusoids

We now show mathematically that the DFT sinusoids are exactly orthogonal. Let

$$s_k(n) \triangleq e^{j\omega_k nT} = e^{j2\pi kn/N} = \left[W_N^k\right]^n, \quad n = 0, 1, 2, \ldots, N-1,$$

denote the kth DFT complex-sinusoid, for $k = 0 : N-1$. Then

$$\begin{aligned} \langle s_k, s_l \rangle &\triangleq \sum_{n=0}^{N-1} s_k(n)\overline{s_l(n)} = \sum_{n=0}^{N-1} e^{j2\pi kn/N} e^{-j2\pi ln/N} \\ &= \sum_{n=0}^{N-1} e^{j2\pi(k-l)n/N} = \frac{1 - e^{j2\pi(k-l)}}{1 - e^{j2\pi(k-l)/N}} \end{aligned}$$

where the last step made use of the closed-form expression for the sum of a geometric series (§6.1). If $k \neq l$, the denominator is nonzero while the numerator is zero. This proves

$$\boxed{s_k \perp s_l, \quad k \neq l.}$$

While we only looked at unit amplitude, zero-phase complex sinusoids, as used by the DFT, it is readily verified that the (nonzero) amplitude and phase have no effect on orthogonality.

6.4 Norm of the DFT Sinusoids

For $k = l$, we follow the previous derivation to the next-to-last step to get

$$\langle s_k, s_k \rangle = \sum_{n=0}^{N-1} e^{j2\pi(k-k)n/N} = N$$

which proves

$$\boxed{\| s_k \| = \sqrt{N}.}$$

6.5 An Orthonormal Sinusoidal Set

We can normalize the DFT sinusoids to obtain an orthonormal set:

$$\tilde{s}_k(n) \stackrel{\Delta}{=} \frac{s_k(n)}{\sqrt{N}} = \frac{e^{j2\pi kn/N}}{\sqrt{N}}$$

The orthonormal sinusoidal basis signals satisfy

$$\langle \tilde{s}_k, \tilde{s}_l \rangle = \begin{cases} 1, & k = l \\ 0, & k \neq l. \end{cases}$$

We call these the *normalized DFT sinusoids*. In §6.10 below, we will project signals onto them to obtain the *normalized DFT* (NDFT).

6.6 The Discrete Fourier Transform (DFT)

Given a signal $x(\cdot) \in \mathbf{C}^N$, its DFT is defined by[3]

$$X(\omega_k) \stackrel{\Delta}{=} \langle x, s_k \rangle \stackrel{\Delta}{=} \sum_{n=0}^{N-1} x(n)\overline{s_k(n)}, \quad k = 0, 1, 2, \ldots, N-1,$$

where

$$s_k(n) \stackrel{\Delta}{=} e^{j\omega_k t_n}, \quad t_n \stackrel{\Delta}{=} nT, \quad \omega_k \stackrel{\Delta}{=} 2\pi\frac{k}{N}f_s, \quad f_s \stackrel{\Delta}{=} \frac{1}{T},$$

or, as it is most often written,

$$\boxed{X(\omega_k) \stackrel{\Delta}{=} \sum_{n=0}^{N-1} x(n)e^{-j\frac{2\pi kn}{N}}, \quad k = 0, 1, 2, \ldots, N-1.}$$

We may also refer to X as the *spectrum* of x, and $X(\omega_k)$ is the kth *sample* of the spectrum at frequency ω_k. Thus, the kth sample $X(\omega_k)$ of the spectrum of x is defined as the inner product of x with the kth DFT sinusoid s_k. This definition is N times the *coefficient of projection* of x onto s_k, *i.e.*,

$$\frac{\langle x, s_k \rangle}{\| s_k \|^2} = \frac{X(\omega_k)}{N}.$$

The projection of x onto s_k is

$$\mathbf{P}_{s_k}(x) = \frac{X(\omega_k)}{N} s_k.$$

[3]As introduced in §1.3 on page 3, the notation $x(\cdot)$ means the *whole signal* $x(n)$, $n = 0, 1, \ldots, N-1$, also written as simply x.

Since the $\{s_k\}$ are orthogonal and span $\mathbf{C}^N$, using the main result of the preceding chapter, we have that the inverse DFT is given by the sum of the projections

$$x = \sum_{k=0}^{N-1} \frac{X(\omega_k)}{N} s_k,$$

or, as we normally write,

$$\boxed{x(n) = \frac{1}{N}\sum_{k=0}^{N-1} X(\omega_k) e^{j\frac{2\pi kn}{N}}, \quad n = 0, 1, \ldots, N-1.}$$

In summary, the DFT is proportional to the set of coefficients of projection onto the sinusoidal basis set, and the inverse DFT is the reconstruction of the original signal as a superposition of its sinusoidal projections. This basic "architecture" extends to all linear orthogonal transforms, including wavelets, Fourier transforms, Fourier series, the discrete-time Fourier transform (DTFT), and certain short-time Fourier transforms (STFT). See Appendix B for some of these.

We have defined the DFT from a geometric signal theory point of view, building on the preceding chapter. See §7.1.1 on page 116 for notation and terminology associated with the DFT.

6.7 Frequencies in the "Cracks"

The DFT is defined only for frequencies $\omega_k = 2\pi k f_s/N$. If we are analyzing one or more periods of an exactly periodic signal, where the period is exactly N samples (or some integer divisor of N), then these really are the only frequencies present in the signal, and the spectrum is actually zero everywhere but at $\omega = \omega_k$, $k \in [0, N-1]$. However, we use the DFT to analyze arbitrary signals from nature. What happens when a frequency ω is present in a signal x that is not one of the DFT-sinusoid frequencies ω_k?

To find out, let's project a length N segment of a sinusoid at an arbitrary frequency ω_x onto the kth DFT sinusoid:

$$\begin{aligned}
x(n) &\triangleq e^{j\omega_x nT} \\
s_k(n) &\triangleq e^{j\omega_k nT} \\
\mathbf{P}_{s_k}(x) &= \frac{\langle x, s_k\rangle}{\langle s_k, s_k\rangle} s_k \triangleq \frac{X(\omega_k)}{N} s_k
\end{aligned}$$

The coefficient of projection is proportional to

$$\begin{aligned} X(\omega_k) &\isdef \langle x, s_k\rangle \isdef \sum_{n=0}^{N-1} x(n)\overline{s_k(n)} \\ &= \sum_{n=0}^{N-1} e^{j\omega_x nT} e^{-j\omega_k nT} = \sum_{n=0}^{N-1} e^{j(\omega_x-\omega_k)nT} = \frac{1-e^{j(\omega_x-\omega_k)NT}}{1-e^{j(\omega_x-\omega_k)T}} \\ &= e^{j(\omega_x-\omega_k)(N-1)T/2}\frac{\sin[(\omega_x-\omega_k)NT/2]}{\sin[(\omega_x-\omega_k)T/2]}, \end{aligned}$$

using the closed-form expression for a geometric series sum once again. As shown in §6.3–§6.4 above, the sum is N if $\omega_k = \omega_x$ and zero at ω_l, for $l \neq k$. However, the sum is *nonzero at all other frequencies* ω_x.

Since we are only looking at N samples, any sinusoidal segment can be projected onto the N DFT sinusoids and be reconstructed exactly by a linear combination of them. Another way to say this is that the DFT sinusoids form a *basis* for $\mathbf{C}^N$, so that any length N signal whatsoever can be expressed as a linear combination of them. Therefore, when analyzing segments of recorded signals, we must interpret what we see accordingly.

The typical way to think about this in practice is to consider the DFT operation as a *digital filter* for each k, whose input is x and whose output is $X(\omega_k)$ at time $n = N - 1$.[4] The *frequency response* of this filter is what we just computed,[5] and its magnitude is

$$|X(\omega_k)| = \left|\frac{\sin[(\omega_x-\omega_k)NT/2]}{\sin[(\omega_x-\omega_k)T/2]}\right|$$

(shown in Fig. 6.3a for $k = N/4$). At all other integer values of k, the frequency response is the same but shifted (circularly) left or right so that the peak is centered on ω_k. The secondary peaks away from ω_k are called *sidelobes* of the DFT response, while the main peak may be called the *main lobe* of the response. Since we are normally most interested in spectra from an audio perspective, the same plot is repeated using a *decibel* vertical scale in Fig. 6.3b[6] (clipped at -60 dB). We see that the sidelobes are really quite high from an audio perspective. Sinusoids with frequencies near $\omega_{k\pm1.5}$, for example, are only attenuated approximately 13 dB in the DFT output $X(\omega_k)$.

We see that $X(\omega_k)$ is sensitive to *all* frequencies between dc and the sampling rate *except* the other DFT-sinusoid frequencies ω_l for $l \neq k$. This

[4] More precisely, $\text{DFT}_k()$ is a length N finite-impulse-response (FIR) digital filter. See §8.3 for related discussion.

[5] We call this the *aliased sinc function* to distinguish it from the *sinc function* $\text{sinc}(x) \isdef \sin(\pi x)/(\pi x)$.

[6] The Matlab code for this figure is given in §I.4.2.

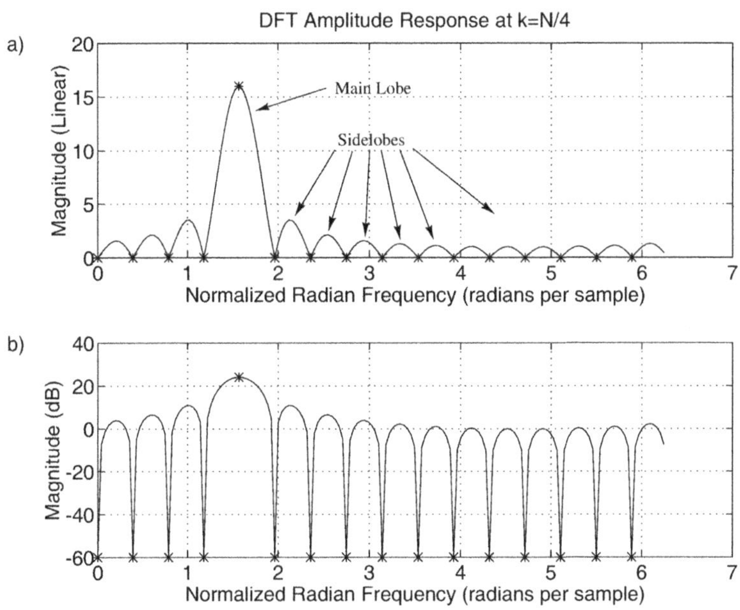

Figure 6.3: Magnitude frequency response of a particular DFT "bin" (where "bin" is defined in §6.8 on page 108). The solid curve shows the relative contribution of arbitrary frequency components to the spectral bin at one-fourth the sampling rate.

is sometimes called *spectral leakage* or *cross-talk* in the spectrum analysis. Again, there is *no leakage* when the signal being analyzed is truly periodic and we can choose N to be exactly a period, or some multiple of a period. Normally, however, this cannot be easily arranged, and spectral leakage can be a problem.

Note that peak spectral leakage is not reduced by increasing N.[7] It can be thought of as being caused by abruptly *truncating* a sinusoid at the beginning and/or end of the N-sample time window. Only the DFT sinusoids are not cut off at the window boundaries. All other frequencies will suffer some truncation distortion, and the spectral content of the abrupt cut-off or turn-on transient can be viewed as the source of the sidelobes. Remember that, as far as the DFT is concerned, the input signal $x(n)$ is the same as its *periodic extension* (more about this in §7.1.2 on page 116). If we repeat N samples of a sinusoid at frequency $\omega \neq \omega_k$ (for any $k \in \mathbf{Z}$), there will be a "glitch" every N samples since the signal is not periodic in N samples. This glitch can be considered a source of new energy over the entire spectrum. See Fig. 8.3 on page 170 for an example waveform.

To reduce spectral leakage (cross-talk from far-away frequencies), we typically use a *window* function, such as a "raised cosine" window, to *taper* the data record gracefully to zero at both endpoints of the window. As a result of the smooth tapering, the *main lobe widens* and the *sidelobes decrease* in the DFT response. Using no window is better viewed as using a *rectangular window* of length N, unless the signal is exactly periodic in N samples. These topics are considered further in Chapter 8.

6.8 Spectral Bin Numbers

Since the kth spectral sample $X(\omega_k)$ is properly regarded as a measure of spectral amplitude over a *range* of frequencies, nominally $k - 1/2$ to $k + 1/2$, this range is sometimes called a *frequency bin* (as in a "storage bin" for spectral energy). The frequency index k is called the *bin number*, and $|X(\omega_k)|^2$ can be regarded as the total energy in the kth bin (see §7.4.9). Similar remarks apply to samples of any bandlimited function; however, the term "bin" is only used in the frequency domain, even though it could be assigned exactly the same meaning mathematically in the time domain.

---

[7]Spectral leakage is essentially equivalent to (*i.e.*, a Fourier dual of) the *Gibb's phenomenon* for truncated Fourier series expansions (see §B.3 on page 210), which some of us studied in high school. As more sinusoids are added to the expansion, the error waveform increases in frequency, and decreases in signal energy, but its peak value does not converge to zero. Instead, in the limit as infinitely many sinusoids are added to the Fourier-series sum, the peak error converges to an *isolated point*. Isolated points have "measure zero" under integration, and therefore have no effect on integrals such as the one which calculates Fourier-series coefficients.

6.9 Fourier Series Special Case

In the very special case of *truly periodic* signals $x(t) = x(t+NT)$, for all $t \in (-\infty, \infty)$, the DFT may be regarded as computing the *Fourier series coefficients* of $x(t)$ from one period of its sampled representation $x(nT)$, $n = 0, 1, \ldots, N-1$. The period of x must be exactly NT seconds for this to work. For the details, see §B.3 on page 210.

6.10 Normalized DFT

A more "theoretically clean" DFT is obtained by projecting onto the *normalized DFT sinusoids* (§6.5)

$$\tilde{s}_k(n) \stackrel{\Delta}{=} \frac{e^{j2\pi kn/N}}{\sqrt{N}}.$$

In this case, the *normalized DFT (NDFT)* of x is

$$\tilde{X}(\omega_k) \stackrel{\Delta}{=} \langle x, \tilde{s}_k \rangle = \frac{1}{\sqrt{N}} \sum_{n=0}^{N-1} x(n) e^{-j2\pi kn/N}$$

which is also precisely the coefficient of projection of x onto $\tilde{s}_k$. The inverse normalized DFT is then more simply

$$x(n) = \sum_{k=0}^{N-1} \tilde{X}(\omega_k)\tilde{s}_k(n) = \frac{1}{\sqrt{N}} \sum_{k=0}^{N-1} \tilde{X}(\omega_k) e^{j2\pi kn/N}.$$

While this definition is much cleaner from a "geometric signal theory" point of view, it is rarely used in practice since it requires slightly more computation than the typical definition. However, note that the only difference between the forward and inverse transforms in this case is the sign of the exponent in the kernel.

It can be said that only the NDFT provides a proper *change of coordinates* from the time-domain (shifted impulse basis signals) to the frequency-domain (DFT sinusoid basis signals). That is, only the NDFT is a pure *rotation* in $\mathbf{C}^N$, preserving both orthogonality and the unit-norm property of the basis functions. The DFT, in contrast, preserves orthogonality, but the norms of the basis functions grow to $\sqrt{N}$. Therefore, in the present context, the DFT coefficients can be considered "denormalized" frequency-domain coordinates.

6.11 The Length 2 DFT

The length 2 DFT is particularly simple, since the basis sinusoids are real:

$$\begin{aligned} \underline{s}_0 &= (1,1) \\ \underline{s}_1 &= (1,-1) \end{aligned}$$

The DFT sinusoid $\underline{s}_0$ is a sampled constant signal, while $\underline{s}_1$ is a sampled sinusoid at half the sampling rate.

Figure 6.4 illustrates the graphical relationships for the length 2 DFT of the signal $\underline{x} = [6, 2]$.

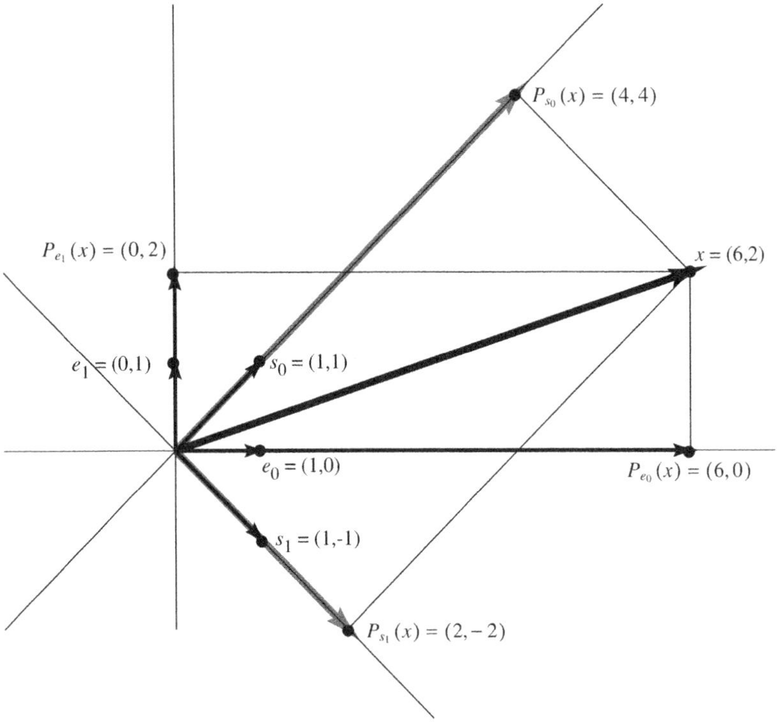

Figure 6.4: Graphical interpretation of the length 2 DFT.

Analytically, we compute the DFT to be

$$\begin{aligned} X(\omega_0) &= \langle \underline{x}, \underline{s}_0 \rangle = 6 \cdot 1 + 2 \cdot 1 = 8 \\ X(\omega_1) &= \langle \underline{x}, \underline{s}_1 \rangle = 6 \cdot 1 + 2 \cdot (-1) = 4 \end{aligned}$$

and the corresponding projections onto the DFT sinusoids are

$$\begin{aligned}
\mathbf{P}_{\underline{s}_0}(\underline{x}) &\triangleq \frac{\langle \underline{x}, \underline{s}_0\rangle}{\langle \underline{s}_0, \underline{s}_0\rangle}\underline{s}_0 = \frac{6\cdot 1+2\cdot 1}{1^2+1^2}\underline{s}_0 = 4\underline{s}_0 = (4,4), \text{ and} \\
\mathbf{P}_{\underline{s}_1}(\underline{x}) &\triangleq \frac{\langle \underline{x}, \underline{s}_1\rangle}{\langle \underline{s}_1, \underline{s}_1\rangle}\underline{s}_1 = \frac{6\cdot 1+2\cdot(-1)}{1^2+(-1)^2}\underline{s}_1 = 2\underline{s}_1 = (2,-2).
\end{aligned}$$

Note the lines of orthogonal projection illustrated in the figure. The "time domain" basis consists of the vectors $\{\underline{e}_0, \underline{e}_1\}$, and the orthogonal projections onto them are simply the coordinate axis projections $(6,0)$ and $(0,2)$. The "frequency domain" basis vectors are $\{\underline{s}_0, \underline{s}_1\}$, and they provide an orthogonal basis set that is rotated 45 degrees relative to the time-domain basis vectors. Projecting orthogonally onto them gives $\mathbf{P}_{\underline{s}_0}(\underline{x}) = (4,4)$ and $\mathbf{P}_{\underline{s}_1}(\underline{x}) = (2,-2)$, respectively. The original signal $\underline{x}$ can be expressed either as the vector sum of its coordinate projections (0,...,x(i),...,0), (a time-domain representation), or as the vector sum of its projections onto the DFT sinusoids (a frequency-domain representation of the time-domain signal $\underline{x}$). Computing the coefficients of projection is essentially "taking the DFT," and constructing $\underline{x}$ as the vector sum of its projections onto the DFT sinusoids amounts to "taking the inverse DFT."

In summary, the oblique coordinates in Fig. 6.4 are interpreted as follows:

$$\begin{aligned}
\underline{x} = (6,2) &= (4,4)+(2,-2) = 4\cdot(1,1)+2\cdot(1,-1) \\
&= 4\cdot \text{dc} + 2\cdot \text{sinusoid-at-half-the-sampling-rate} \\
&= \frac{X(\omega_0)}{\|\underline{s}_0\|^2}\underline{s}_0 + \frac{X(\omega_1)}{\|\underline{s}_1\|^2}\underline{s}_1
\end{aligned}$$

6.12 Matrix Formulation of the DFT

The DFT can be formulated as a complex matrix multiply, as we show in this section. (This section can be omitted without affecting what follows.) For basic definitions regarding matrices, see Appendix H.

The DFT consists of inner products of the input signal $\underline{x}$ with sampled complex sinusoidal sections $\underline{s}_k$:

$$X(\omega_k) \triangleq \langle \underline{x}, \underline{s}_k\rangle \triangleq \sum_{n=0}^{N-1} \underline{x}(n)e^{-j2\pi nk/N}, \quad k = 0,1,2,\ldots,N-1$$

By collecting the DFT output samples into a column vector, we have

$$\underbrace{\begin{bmatrix} X(\omega_0) \\ X(\omega_1) \\ X(\omega_2) \\ \vdots \\ X(\omega_{N-1}) \end{bmatrix}}_{\underline{X}} = \underbrace{\begin{bmatrix} \overline{s_0(0)} & \overline{s_0(1)} & \cdots & \overline{s_0(N-1)} \\ \overline{s_1(0)} & \overline{s_1(1)} & \cdots & \overline{s_1(N-1)} \\ \overline{s_2(0)} & \overline{s_2(1)} & \cdots & \overline{s_2(N-1)} \\ \vdots & \vdots & \vdots & \vdots \\ \overline{s_{N-1}(0)} & \overline{s_{N-1}(1)} & \cdots & \overline{s_{N-1}(N-1)} \end{bmatrix}}_{\mathbf{S}_N^*} \underbrace{\begin{bmatrix} x(0) \\ x(1) \\ x(2) \\ \vdots \\ x(N-1) \end{bmatrix}}_{\underline{x}}$$

or

$$\underline{X} = \mathbf{S}_N^* \underline{x} = \begin{bmatrix} \underline{s}_0^* \\ \vdots \\ \underline{s}_{N-1}^* \end{bmatrix} \underline{x} = \begin{bmatrix} \langle \underline{x}, \underline{s}_0 \rangle \\ \vdots \\ \langle \underline{x}, \underline{s}_{N-1} \rangle \end{bmatrix}$$

where $\mathbf{S}_N^*$ denotes the *DFT matrix* $\mathbf{S}_N^*[k,n] \triangleq W_N^{-kn} \triangleq e^{-j2\pi kn/N}$, *i.e.*,

$$\begin{aligned} \mathbf{S}_N^* &\triangleq \begin{bmatrix} \underline{s}_0 & \underline{s}_1 & \cdots & \underline{s}_{N-1} \end{bmatrix}^* \\ &\triangleq \begin{bmatrix} \overline{s_0(0)} & \overline{s_0(1)} & \cdots & \overline{s_0(N-1)} \\ \overline{s_1(0)} & \overline{s_1(1)} & \cdots & \overline{s_1(N-1)} \\ \overline{s_2(0)} & \overline{s_2(1)} & \cdots & \overline{s_2(N-1)} \\ \vdots & \vdots & \vdots & \vdots \\ \overline{s_{N-1}(0)} & \overline{s_{N-1}(1)} & \cdots & \overline{s_{N-1}(N-1)} \end{bmatrix} \\ &\triangleq \begin{bmatrix} 1 & 1 & 1 & \cdots & 1 \\ 1 & e^{-j2\pi/N} & e^{-j4\pi/N} & \cdots & e^{-j2\pi(N-1)/N} \\ 1 & e^{-j4\pi/N} & e^{-j8\pi/N} & \cdots & e^{-j2\pi 2(N-1)/N} \\ \vdots & \vdots & \vdots & \vdots & \vdots \\ 1 & e^{-j2\pi(N-1)/N} & e^{-j2\pi 2(N-1)/N} & \cdots & e^{-j2\pi(N-1)(N-1)/N} \end{bmatrix}. \end{aligned}$$

The notation $A^* \triangleq \overline{A^{\mathrm{T}}}$ denotes the *Hermitian transpose* of the complex matrix A (transposition and complex conjugation).

Note that the kth column of $\mathbf{S}_N$ is the kth DFT sinusoid, so that the kth row of the DFT matrix $\mathbf{S}_N^*$ is the complex-conjugate of the kth DFT sinusoid. Therefore, multiplying the DFT matrix times a signal vector $\underline{x}$ produces a column-vector $\underline{X} = \mathbf{S}_N^*\underline{x}$ in which the kth element $\underline{X}[k]$ is the inner product of the kth DFT sinusoid with $\underline{x}$, or $\underline{X}[k] = \underline{s}_k^*\underline{x} = \langle \underline{x}, s_k \rangle$, as expected.

Computation of the DFT matrix in Matlab is illustrated in §I.4.3 on page 286.

The *inverse DFT matrix* is simply $\mathbf{S}_N/N$. That is, we can perform the inverse DFT operation as

$$\underline{x} = \frac{1}{N}\mathbf{S}_N\underline{X}. \tag{6.1}$$

Since the forward DFT is $\underline{X} = \mathbf{S}_N^*\underline{x}$, substituting $\underline{x}$ from Eq. (6.1) into the forward DFT leads quickly to the conclusion that

$$\mathbf{S}_N^*\mathbf{S}_N = N \cdot \mathbf{I}. \tag{6.2}$$

This equation succinctly states that the *columns of* $\mathbf{S}_N$ *are orthogonal*, which, of course, we already knew. *I.e.*, $\langle \underline{s}_k, \underline{s}_n\rangle = 0$ for $k \neq n$, and $\langle \underline{s}_k, \underline{s}_k\rangle = N$:

$$\begin{aligned}\mathbf{S}_N^*\mathbf{S}_N &= \left[\begin{array}{cccc}\langle \underline{s}_0, \underline{s}_0\rangle & \langle \underline{s}_0, \underline{s}_1\rangle & \cdots & \langle \underline{s}_0, \underline{s}_{N-1}\rangle \\ \langle \underline{s}_1, \underline{s}_0\rangle & \langle \underline{s}_1, \underline{s}_1\rangle & \cdots & \langle \underline{s}_1, \underline{s}_{N-1}\rangle \\ \langle \underline{s}_2, \underline{s}_0\rangle & \langle \underline{s}_2, \underline{s}_1\rangle & \cdots & \langle \underline{s}_2, \underline{s}_{N-1}\rangle \\ \vdots & \vdots & \vdots & \vdots \\ \langle \underline{s}_{N-1}, \underline{s}_0\rangle & \langle \underline{s}_{N-1}, \underline{s}_1\rangle & \cdots & \langle \underline{s}_{N-1}, \underline{s}_{N-1}\rangle\end{array}\right] \\ &= \left[\begin{array}{ccccc}N & 0 & 0 & \cdots & 0 \\ 0 & N & 0 & \cdots & 0 \\ 0 & 0 & N & \cdots & 0 \\ \vdots & \vdots & \vdots & \vdots & \vdots \\ 0 & 0 & 0 & \cdots & N\end{array}\right] = N \cdot \mathbf{I}.\end{aligned}$$

The *normalized DFT matrix* is given by

$$\tilde{\mathbf{S}}_N^* \triangleq \frac{1}{\sqrt{N}}\mathbf{S}_N^*$$

and the corresponding normalized *inverse* DFT matrix is simply $\tilde{\mathbf{S}}_N$, so that Eq. (6.2) becomes

$$\tilde{\mathbf{S}}_N^*\tilde{\mathbf{S}}_N = \mathbf{I}.$$

This implies that the columns of $\tilde{\mathbf{S}}_N$ are *orthonormal*. Such a complex matrix is said to be *unitary*.

When a *real* matrix $\mathbf{Q}$ satisfies $\mathbf{Q}^T\mathbf{Q} = \mathbf{I}$, then $\mathbf{Q}$ is said to be *orthogonal*. "Unitary" is the generalization of "orthogonal" to complex matrices.

6.13 DFT Problems

1. Show that if two length N sampled sinusoids $s_k(n)$ and $s_l(n)$ are orthogonal, *i.e.*, $s_k \perp s_l$, then $\mathcal{A}s_k \perp \mathcal{B}s_l$ for all complex constants $\mathcal{A}$ and $\mathcal{B}$. That is, the orthogonality of two sinusoids is independent of their phases and (nonzero) amplitudes.

2. Find the length $N = 8$ DFTs of the following sequences:

   (a) $x_1 = [1, 0, 0, 0, 0, 0, 0, 0]$ (the "impulse" signal)

   (b) $x_2 = [0, 1, 0, 0, 0, 0, 0, 0]$ (delayed impulse)

   (c) $x_3 = [1, 1, 1, 1, 1, 1, 1, 1]$ ("dc")

   (d) $x_4 = [2, 1, 0, 0, 0, 0, 0, 1]$

   (e) Express $X_4(k)$ in terms of $X_1(k)$ and $X_2(k)$, where $X_i(k)$ denotes the kth bin of the DFT of x_i.

   (f) Sketch the magnitude spectrum and phase spectrum for the sequence $x = [1, 2, 1, 0, 0, \ldots]$. [Note that now $N \to \infty$, so $X = \text{DTFT}(x)$.] Label all extreme values.

3. Show how to construct the DFT of a length $2N$ *real-valued* signal from a length N complex DFT. Start by placing the first N points in the real part of the DFT input, and the remaining N points in the imaginary part.

4. Find the length N DFT for the following signals $x \in \mathbf{R}^N$

   (a) Impulse train: $x(n) = \begin{cases} 1, & n \text{ even} \\ 0, & n \text{ odd} \end{cases}$

   (b) Sampled complex sinusoid: $x(n) = e^{j2\pi nl/N}$, l any integer

5. A *rotation* can be defined as a linear transformation which preserves angles. Show that a real matrix R corresponds to a rotation in $\mathbf{R}^N$ if and only if $R^T R = I$. [Hint: recall from §5.9.6 that the cosine of the angle between two real vectors can be defined as a normalized inner product of the vectors.]

6. Generalize the previous problem to *complex unitary matrices* R.

Chapter 7

Fourier Theorems for the DFT

This chapter derives various *Fourier theorems* for the case of the DFT. Included are symmetry relations, the shift theorem, convolution theorem, correlation theorem, power theorem, and theorems pertaining to interpolation and downsampling. Applications related to certain theorems are outlined, including linear time-invariant filtering, sampling rate conversion, and statistical signal processing.

7.1 The DFT and its Inverse Restated

Let $x(n), n = 0, 1, 2, \ldots, N-1$, denote an N-sample complex sequence, *i.e.*, $x \in \mathbf{C}^N$. Then the *spectrum* of x is defined by the *Discrete Fourier Transform (DFT)*:

$$\boxed{X(k) \triangleq \sum_{n=0}^{N-1} x(n)e^{-j2\pi nk/N}, \quad k = 0, 1, 2, \ldots, N-1}$$

The *inverse DFT (IDFT)* is defined by

$$\boxed{x(n) = \frac{1}{N}\sum_{k=0}^{N-1} X(k)e^{j2\pi nk/N}, \quad n = 0, 1, 2, \ldots, N-1.}$$

In this chapter, we will omit mention of an explicit sampling interval $T = 1/f_s$, as this is most typical in the digital signal processing literature. It is often said that the sampling frequency is $f_s = 1$. In this case, a radian frequency $\omega_k \stackrel{\Delta}{=} 2\pi k/N$ is in units of "radians per sample." Elsewhere in this book, ω_k usually means "radians per *second*." (Of course, there's no difference when the sampling rate is really 1.) Another term we use in connection with the $f_s = 1$ convention is *normalized frequency*: All normalized radian frequencies lie in the range $[-\pi, \pi)$, and all normalized frequencies in Hz lie in the range $[-0.5, 0.5)$.[1] Note that physical units of seconds and Hertz can be reintroduced by the substitution

$$e^{j2\pi nk/N} = e^{j2\pi k(f_s/N)nT} \stackrel{\Delta}{=} e^{j\omega_k t_n}.$$

7.1.1 Notation and Terminology

If X is the DFT of x, we say that x and X form a *transform pair* and write

$$\boxed{x \longleftrightarrow X} \quad \text{("}x\text{ corresponds to }X\text{")}.$$

Another notation we'll use is

$$\begin{aligned} \text{DFT}(x) &\stackrel{\Delta}{=} X, \quad \text{and} \\ \text{DFT}_k(x) &\stackrel{\Delta}{=} X(k). \end{aligned}$$

If we need to indicate the length of the DFT explicitly, we will write $\text{DFT}_N(x) = X$ and $\text{DFT}_{N,k}(x) = X(k)$. As we've already seen, *time-domain* signals are consistently denoted using *lowercase* symbols such as "$x(n)$," while *frequency-domain* signals (spectra), are denoted in *uppercase* ("$X(\omega_k)$").

7.1.2 Modulo Indexing, Periodic Extension

The DFT sinusoids $s_k(n) \stackrel{\Delta}{=} e^{j\omega_k n}$ are all *periodic* having periods which divide N. That is, $s_k(n + mN) = s_k(n)$ for any integer m. Since a length N signal x can be expressed as a linear combination of the DFT sinusoids in the time domain,

$$x(n) = \frac{1}{N} \sum_k X(k) s_k(n),$$

it follows that the "automatic" definition of $x(n)$ beyond the range $[0, N-1]$ is *periodic extension, i.e.*, $x(n + mN) \stackrel{\Delta}{=} x(n)$ for every integer m.

Moreover, the DFT also repeats naturally every N samples, since

$$X(k + mN) \stackrel{\Delta}{=} \langle x, s_{k+mN} \rangle = \langle x, s_k \rangle = X(k)$$

[1]The notation $[a, b)$ denotes the *half-open interval* on the real line from a to b. Thus the interval includes a but not b.

because $s_{k+mN}(n) = e^{j2\pi n(k+mN)/N} = e^{j2\pi nk/N}e^{j2\pi nm} = s_k(n)$. (The DFT sinusoids behave identically as functions of n and k.) Accordingly, for purposes of DFT studies, we may define *all* signals in $\mathbf{C}^N$ as being single periods from an infinitely long periodic signal with period N samples:

Definition (Periodic Extension): For any signal $x \in \mathbf{C}^N$, we define

$$x(n+mN) \stackrel{\Delta}{=} x(n)$$

for every integer m.

As a result of this convention, all indexing of signals and spectra[2] can be interpreted *modulo* N, and we may write $x(n\,(\text{mod}\,N))$ to emphasize this. Formally, "$n\,(\text{mod}\,N)$" is defined as $n - mN$ with m chosen to give $n - mN$ in the range $[0, N-1]$.

As an example, when indexing a spectrum X, we have that $X(N) = X(0)$ which can be interpreted physically as saying that the sampling rate is the same frequency as dc for discrete time signals. Periodic extension in the time domain implies that the signal input to the DFT is mathematically treated as being *samples of one period of a periodic signal*, with the period being exactly NT seconds (N samples). The corresponding assumption in the frequency domain is that the spectrum is *exactly zero between frequency samples* ω_k. It is also possible to adopt the point of view that the time-domain signal $x(n)$ consists of N samples preceded and followed by *zeros*. In that case, the spectrum would be *nonzero* between spectral samples ω_k, and the spectrum between samples would be reconstructed by means of *bandlimited interpolation* [69].

7.2 Signal Operators

It will be convenient in the Fourier theorems of §7.4 to make use of the following *signal operator* definitions.

7.2.1 Operator Notation

In this book, an *operator* is defined as a *signal-valued function of a signal*. Thus, for the space of length N complex sequences, an operator Op is a mapping from $\mathbf{C}^N$ to $\mathbf{C}^N$:

$$\text{Op}(x) \in \mathbf{C}^N \;\forall x \in \mathbf{C}^N$$

[2] A spectrum is mathematically identical to a signal, since both are just sequences of N complex numbers. However, for clarity, we generally use "signal" when the sequence index is considered a time index, and "spectrum" when the index is associated with successive frequency samples.

An example is the DFT operator:

$$\text{DFT}(x) = X$$

The argument to an operator is always an entire signal. However, its output may be subscripted to obtain a specific sample, *e.g.*,

$$\text{DFT}_k(x) = X(k).$$

Some operators require one or more *parameters* affecting their definition. For example the *shift operator* (defined in §7.2.3 below) requires a *shift amount* $\Delta \in \mathbf{Z}$:[3]

$$\text{SHIFT}_{\Delta,n}(x) \stackrel{\Delta}{=} x(n-\Delta)$$

A time or frequency index, if present, will always be the last subscript. Thus, the signal $\text{SHIFT}_\Delta(x)$ is obtained from x by shifting it Δ samples.

Note that operator notation is *not standard* in the field of digital signal processing. It can be regarded as being influenced by the field of computer science. In the Fourier theorems below, both operator and conventional signal-processing notations are provided. In the author's opinion, operator notation is consistently clearer, allowing powerful expressions to be written naturally in one line (*e.g.*, see Eq. (7.8) on page 162), and it is much closer to how things look in a readable computer program (such as in the matlab language).

7.2.2 Flip Operator

We define the *flip operator* by

$$\text{FLIP}_n(x) \stackrel{\Delta}{=} x(-n), \tag{7.1}$$

for all sample indices $n \in \mathbf{Z}$. By modulo indexing, $x(-n)$ is the same as $x(N-n)$. The FLIP() operator reverses the order of samples 1 through $N-1$ of a sequence, leaving sample 0 alone, as shown in Fig. 7.1a. Thanks to modulo indexing, it can also be viewed as "flipping" the sequence about the time 0, as shown in Fig. 7.1b. The interpretation of Fig. 7.1b is usually the one we want, and the FLIP operator is usually thought of as "time reversal" when applied to a signal x or "frequency reversal" when applied to a spectrum X.

7.2.3 Shift Operator

The *shift operator* is defined by

$$\text{SHIFT}_{\Delta,n}(x) \stackrel{\Delta}{=} x(n-\Delta), \quad \Delta \in \mathbf{Z},$$

[3]The set of all integers is denoted $\mathbf{Z}$.

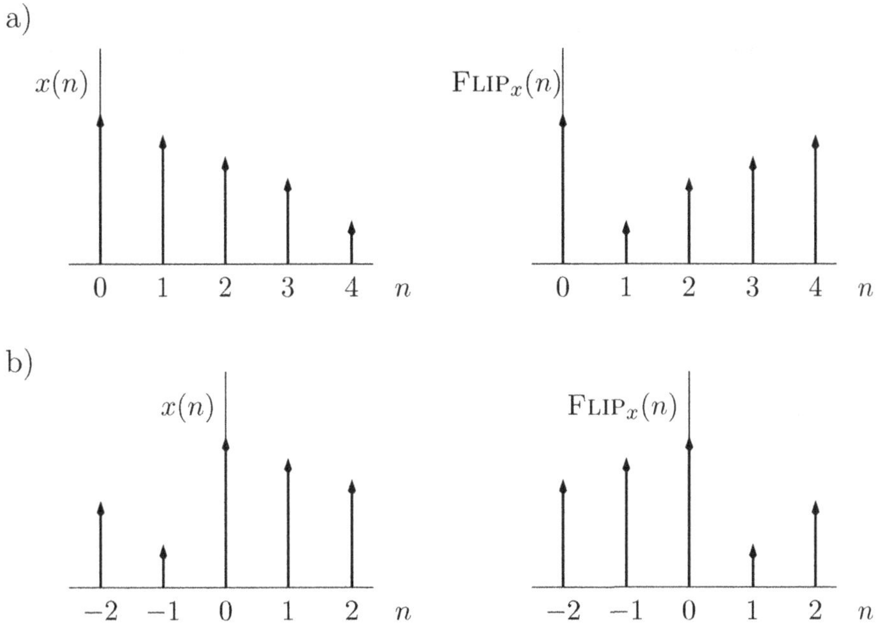

Figure 7.1: Illustration of x and $\text{FLIP}(x)$ for $N = 5$ for the two main domain conventions: a) $n \in [0, N-1]$. b) $n \in [-(N-1)/2, (N-1)/2]$.

and $\text{SHIFT}_\Delta(x)$ denotes the entire shifted signal. Note that since indexing is modulo N, the shift is *circular* (or "cyclic"). However, we normally use it to represent *time delay* by Δ samples. We often use the shift operator in conjunction with *zero padding* (appending zeros to the signal x, §7.2.7) in order to avoid the "wrap-around" associated with a circular shift.

Figure 7.2 illustrates successive one-sample delays of a periodic signal having first period given by $[0, 1, 2, 3, 4]$.

Examples

- $\text{SHIFT}_1([1, 0, 0, 0]) = [0, 1, 0, 0]$ (an impulse delayed one sample).
- $\text{SHIFT}_1([1, 2, 3, 4]) = [4, 1, 2, 3]$ (a circular shift example).
- $\text{SHIFT}_{-2}([1, 0, 0, 0]) = [0, 0, 1, 0]$ (another circular shift example).

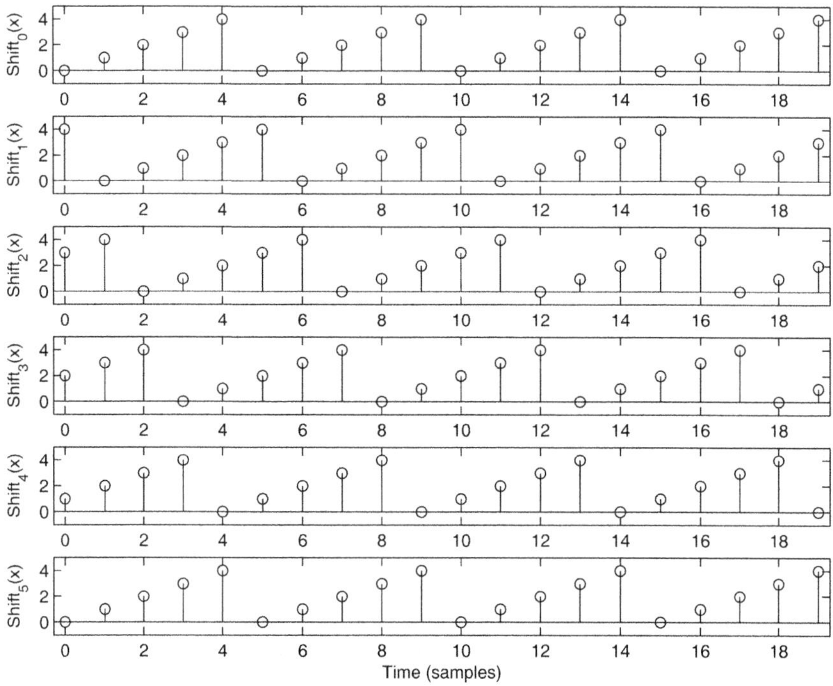

Figure 7.2: Successive one-sample shifts of a sampled periodic sawtooth waveform having first period $[0, 1, 2, 3, 4]$.

7.2.4 Convolution

The *convolution* of two signals x and y in $\mathbf{C}^N$ may be denoted "$x \circledast y$" and defined by

$$\boxed{(x \circledast y)_n \triangleq \sum_{m=0}^{N-1} x(m)y(n-m)}$$

Note that this is *circular convolution* (or "cyclic" convolution).[4] The importance of convolution in *linear systems theory* is discussed in §8.3.

Cyclic convolution can be expressed in terms of previously defined operators as

$$y(n) \triangleq (x \circledast h)_n \triangleq \sum_{m=0}^{N-1} x(m)h(n-m) = \langle x, \text{SHIFT}_n(\text{FLIP}(h))\rangle \quad (h \text{ real})$$

where $x, y \in \mathbf{C}^N$ and $h \in \mathbf{R}^N$. This expression suggests *graphical convolution*, discussed below in §7.2.4.

Commutativity of Convolution

Convolution (cyclic or acyclic) is *commutative*, *i.e.*,

$$\boxed{x \circledast y = y \circledast x.}$$

Proof:

$$\begin{aligned}
(x \circledast y)_n &\triangleq \sum_{m=0}^{N-1} x(m)y(n-m) = \sum_{l=n}^{n-(N-1)} x(n-l)y(l) \\
&= \sum_{l=0}^{N-1} y(l)x(n-l) \\
&\triangleq (y \circledast x)_n
\end{aligned}$$

In the first step we made the change of summation variable $l \triangleq n - m$, and in the second step, we made use of the fact that any sum over all N terms is equivalent to a sum from 0 to $N-1$.

[4]To simulate acyclic convolution using cyclic convolution, as is appropriate for the simulation of sampled continuous-time systems, sufficient *zero padding* is used so that nonzero samples do not "wrap around" as a result of the shifting of y in the definition of convolution. Zero padding is discussed later in this chapter (§7.2.7).

Convolution as a Filtering Operation

In a convolution of two signals $x \circledast y$, where both x and y are signals of length N (real or complex), we may interpret either x or y as a *filter* that operates on the other signal which is in turn interpreted as the filter's "input signal".[5] Let $h \in \mathbf{C}^N$ denote a length N signal that is interpreted as a filter. Then given any input signal $x \in \mathbf{C}^N$, the filter output signal $y \in \mathbf{C}^N$ may be defined as the *cyclic convolution* of x and h:

$$y = h \circledast x = x \circledast h$$

Because the convolution is cyclic, with x and h chosen from the set of (periodically extended) vectors of length N, $h(n)$ is most precisely viewed as the *impulse-train-response* of the associated filter at time n. Specifically, the impulse-train response $h \in \mathbf{C}^N$ is the response of the filter to the *impulse-train signal* $\delta \triangleq [1, 0, \dots, 0] \in \mathbf{R}^N$, which, by periodic extension, is equal to

$$\delta(n) = \begin{cases} 1, & n = 0 \pmod{N} \\ 0, & n \neq 0 \pmod{N}. \end{cases}$$

Thus, N is the *period* of the impulse-train in samples—there is an "impulse" (a '1') every N samples. Neglecting the assumed periodic extension of all signals in $\mathbf{C}^N$, we may refer to δ more simply as the *impulse signal*, and h as the *impulse response* (as opposed to impulse-*train* response). In contrast, for the DTFT (§B.1 on page 208), in which the discrete-time axis is infinitely long, the impulse signal $\delta(n)$ is defined as

$$\delta(n) \triangleq \begin{cases} 1, & n = 0 \\ 0, & n \neq 0 \end{cases}$$

and no periodic extension arises.

As discussed below (§7.2.7), one may embed *acyclic* convolution within a larger cyclic convolution. In this way, real-world systems may be simulated using fast DFT convolutions (see Appendix A for more on fast convolution algorithms).

Note that only linear, time-invariant (LTI) filters can be completely represented by their impulse response (the filter output in response to an impulse at time 0). The convolution representation of LTI digital filters is fully discussed in Book II [65] of the music signal processing book series (in which this is Book I).

[5]See §8.3 on page 185 for an introduction to the digital-filter point of view.

Convolution Example 1: Smoothing a Rectangular Pulse

Figure 7.3 illustrates convolution of

$$x = [0, 0, 0, 0, 1, 1, 1, 1, 1, 1, 0, 0, 0, 0]$$

with

$$h = \left[\frac{1}{3}, \frac{1}{3}, \frac{1}{3}, 0, 0, 0, 0, 0, 0, 0, 0, 0, 0, 0\right]$$

to get

$$y = x \circledast h = \left[0, 0, 0, 0, \frac{1}{3}, \frac{2}{3}, 1, 1, 1, 1, \frac{2}{3}, \frac{1}{3}, 0, 0\right] \tag{7.2}$$

as graphed in Fig. 7.3(c). In this case, h can be viewed as a "moving three-point average" filter. Note how the corners of the rectangular pulse are "smoothed" by the three-point filter. Also note that the pulse is smeared to the "right" (forward in time) because the filter impulse response starts at time zero. Such a filter is said to be *causal* (see [65] for details). By shifting the impulse response left one sample to get

$$h = \left[\frac{1}{3}, \frac{1}{3}, 0, 0, 0, 0, 0, 0, 0, 0, 0, 0, 0, \frac{1}{3}\right]$$

(in which case $\text{FLIP}(h) = h$), we obtain a noncausal filter h which is symmetric about time zero so that the input signal is smoothed "in place" with no added delay (imagine Fig. 7.3(c) shifted left one sample, in which case the input pulse edges align with the midpoint of the rise and fall in the output signal).

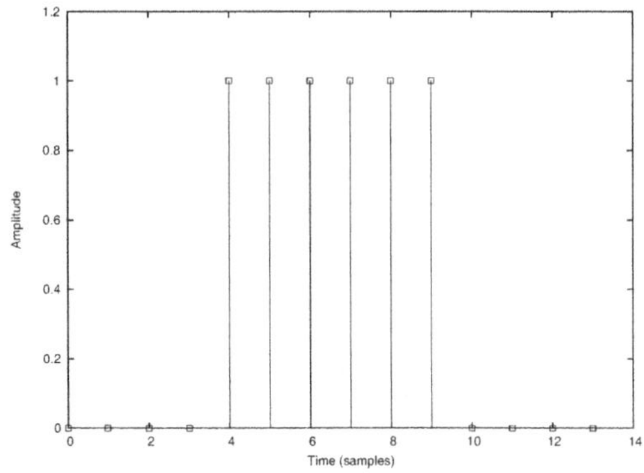

(a) Filter input signal $x(n)$.

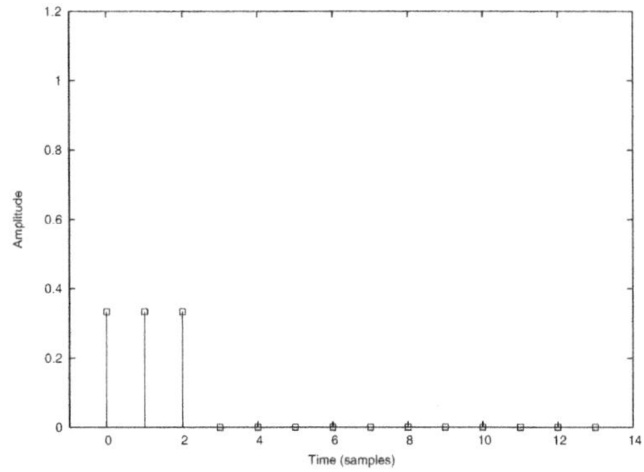

(b) Filter impulse response $h(n)$.

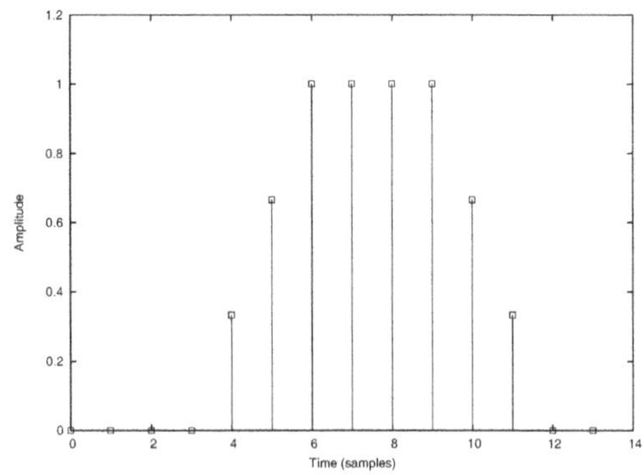

(c) Filter output signal $y(n)$.

Figure 7.3: Illustration of the convolution of a rectangular pulse $x = [0, 0, 0, 0, 1, 1, 1, 1, 1, 1, 0, 0, 0, 0]$ and the impulse response of an "averaging filter" $h = [1/3, 1/3, 1/3, 0, 0, 0, 0, 0, 0, 0, 0, 0, 0, 0]$ $(N = 14)$.

Convolution Example 2: ADSR Envelope

In this example, the input signal is a sequence of two rectangular pulses, creating a piecewise constant function, depicted in Fig. 7.4(a). The filter impulse response, shown in Fig. 7.4(b), is a truncated exponential.[6]

In this example, h is again a causal smoothing-filter impulse response, and we could call it a "moving weighted average", in which the weighting is exponential into the past. The discontinuous steps in the input become exponential "asymptotes" in the output which are approached exponentially. The overall appearance of the output signal resembles what is called an *attack, decay, release, and sustain envelope*, or *ADSR envelope* for short. In a practical ADSR envelope, the time-constants for attack, decay, and release may be set independently. In this example, there is only one time constant, that of h. The two constant levels in the input signal may be called the *attack level* and the *sustain level*, respectively. Thus, the envelope approaches the attack level at the attack rate (where the "rate" may be defined as the reciprocal of the time constant), it next approaches the sustain level at the "decay rate", and finally, it approaches zero at the "release rate". These envelope parameters are commonly used in analog synthesizers and their digital descendants, so-called *virtual analog* synthesizers. Such an ADSR envelope is typically used to multiply the output of a waveform oscillator such as a sawtooth or pulse-train oscillator. For more on virtual analog synthesis, see, for example, [75, 74].

Convolution Example 3: Matched Filtering

Figure 7.5 illustrates convolution of

$$\begin{aligned} y &= [1,1,1,1,0,0,0,0] \\ h &= [1,0,0,0,0,1,1,1] \end{aligned}$$

to get

$$y \circledast h = [4,3,2,1,0,1,2,3]. \qquad (7.3)$$

For example, y could be a "rectangularly windowed signal, zero-padded by a factor of 2," where the signal happened to be dc (all 1s). For the convolution, we need

$$\text{FLIP}(h) = [1,1,1,1,0,0,0,0]$$

[6]Normally, in practice, a *first-order recursive filter* would be used to provide such an exponential impulse response very efficiently in hardware or software (see Book II [65] of this series for details). However, the impulse response of any linear, time-invariant filter can be recorded and used for implementation via convolution. The only catch is that recursive filters generally have *infinitely long* impulse responses (true exponential decays). Therefore, it is necessary to truncate the impulse response when it decays enough that the remainder is negligible.

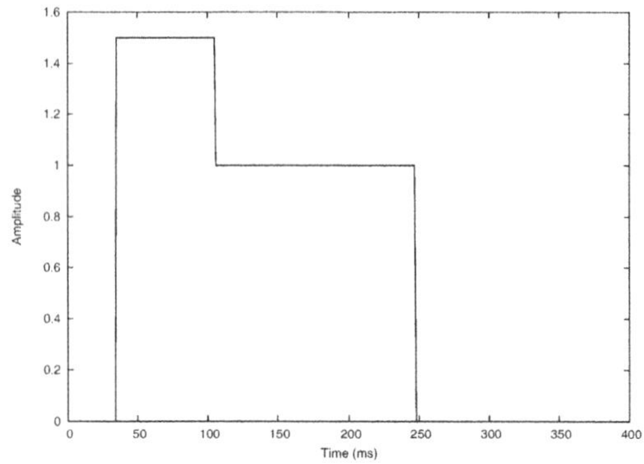

(a) Filter input signal $x(n)$.

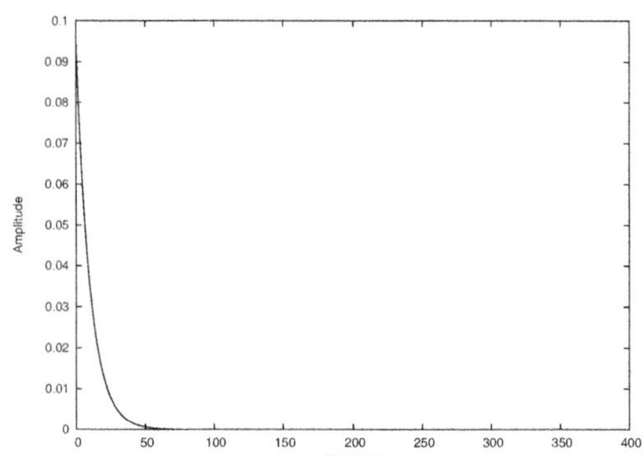

(b) Filter impulse response $h(n)$.

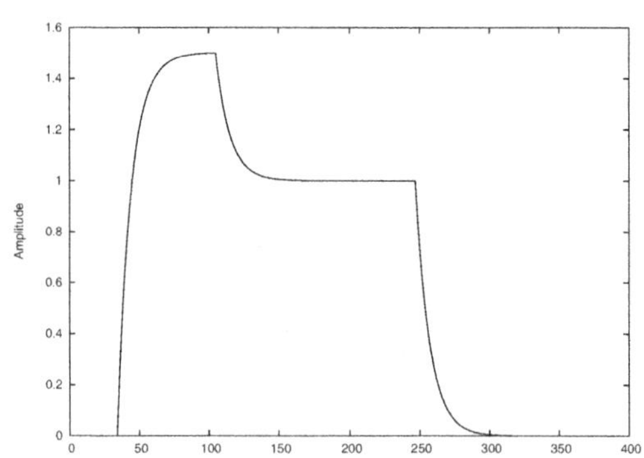

(c) Filter output signal $y(n)$.

Figure 7.4: Illustration of the convolution of two rectangular pulses with a truncated exponential impulse response.

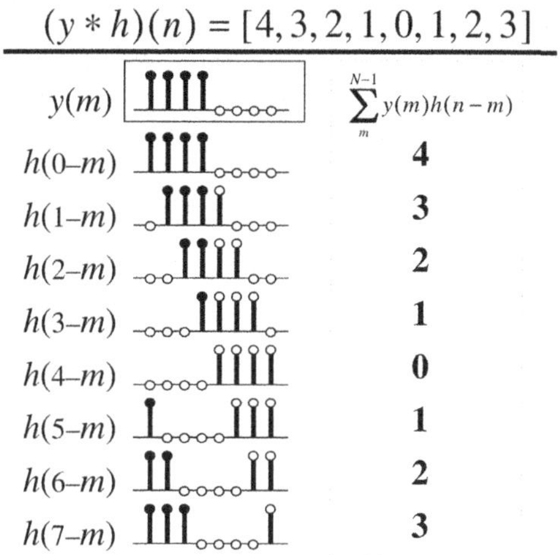

Figure 7.5: Illustration of convolution of $y = [1,1,1,1,0,0,0,0]$ and "matched filter" $h = \text{FLIP}(y) = [1,0,0,0,0,1,1,1]$ ($N = 8$).

which is the same as y. When $h = \text{FLIP}(y)$, we say that h is a *matched filter* for y.[7] In this case, h is matched to look for a "dc component," and also zero-padded by a factor of 2. The zero-padding serves to simulate acyclic convolution using circular convolution. Note from Eq. (7.3) that the maximum is obtained in the convolution output at time 0. This peak (the largest possible if all input signals are limited to $[-1, 1]$ in magnitude), indicates the matched filter has "found" the dc signal starting at time 0. This peak would persist in the presence of some amount of noise and/or interference from other signals. Thus, matched filtering is useful for detecting known signals in the presence of noise and/or interference [32].

Graphical Convolution

As mentioned above, cyclic convolution can be written as

$$y(n) \stackrel{\Delta}{=} (x \circledast h)_n \stackrel{\Delta}{=} \sum_{m=0}^{N-1} x(m)h(n-m) = \langle x, \text{SHIFT}_n(\text{FLIP}(h))\rangle \qquad (h \text{ real})$$

where $x, y \in \mathbf{C}^N$ and $h \in \mathbf{R}^N$. It is instructive to interpret this expression *graphically*, as depicted in Fig. 7.5 above. The convolution result at time $n = 0$ is the inner product of x and $\text{FLIP}(h)$, or $y(0) = \langle x, \text{FLIP}(h)\rangle$. For the next time instant, $n = 1$, we shift $\text{FLIP}(h)$ one

[7] Matched filtering is briefly discussed in §8.4.

sample to the right and repeat the inner product operation to obtain $y(1) = \langle x, \text{SHIFT}_1(\text{FLIP}(h))\rangle$, and so on. To capture the cyclic nature of the convolution, x and $\text{SHIFT}_n(\text{FLIP}(h))$ can be imagined plotted on a *cylinder*. Thus, Fig. 7.5 shows the cylinder after being "cut" along the vertical line between $n = N - 1$ and $n = 0$ and "unrolled" to lay flat.

Polynomial Multiplication

Note that when you multiply two polynomials together, their coefficients are *convolved*. To see this, let $p(x)$ denote the mth-order polynomial

$$p(x) = p_0 + p_1 x + p_2 x^2 + \cdots + p_m x^m$$

with coefficients p_i, and let $q(x)$ denote the nth-order polynomial

$$q(x) = q_0 + q_1 x + q_2 x^2 + \cdots + q_n x^n$$

with coefficients q_i. Then we have [1]

$$\begin{aligned}
p(x)q(x) &= p_0q_0 + (p_0q_1 + p_1q_0)x + (p_0q_2 + p_1q_1 + p_2q_0)x^2 \\
&\quad + (p_0q_3 + p_1q_2 + p_2q_1 + p_3q_0)x^3 \\
&\quad + (p_0q_4 + p_1q_3 + p_2q_2 + p_3q_1 + p_4q_0)x^4 + \cdots \\
&\quad + (p_0q_{n+m} + p_1q_{n+m-1} + p_2q_{n+m-2} \\
&\qquad + p_{n+m-1}q_1 + p_{n+m}q_0)x^{n+m}.
\end{aligned}$$

Denoting $p(x)q(x)$ by

$$r(x) \triangleq p(x)q(x) = r_0 + r_1 x + r_2 x^2 + \cdots + r_{m+n} x^{m+n},$$

we have that the ith coefficient can be expressed as

$$\begin{aligned}
r_i &= p_0q_i + p_1q_{i-1} + p_2q_{i-2} + \cdots + p_{i-1}q_1 + p_iq_0 \\
&= \sum_{j=0}^{i} p_j q_{i-j} = \sum_{j=-\infty}^{\infty} p_j q_{i-j} \\
&\triangleq (p \circledast q)(i),
\end{aligned}$$

where p_i and q_i are doubly infinite sequences, defined as zero for $i < 0$ and $i > m, n$, respectively.

Multiplication of Decimal Numbers

Since decimal numbers are implicitly just polynomials in the powers of 10, *e.g.*,

$$3819 = 3 \cdot 10^3 + 8 \cdot 10^2 + 1 \cdot 10^1 + 9 \cdot 10^0,$$

it follows that *multiplying two numbers convolves their digits.* The only twist is that, unlike normal polynomial multiplication, we have *carries.* That is, when a convolution result (output digit) exceeds 10, we subtract 10 from the result and add 1 to the digit in the next higher place.

7.2.5 Correlation

The *correlation operator* for two signals x and y in $\mathbf{C}^N$ is defined as

$$\boxed{(x \star y)_n \triangleq \sum_{m=0}^{N-1} \overline{x(m)} y(m+n)}$$

We may interpret the correlation operator as

$$(x \star y)_n = \langle \text{SHIFT}_{-n}(y), x \rangle$$

which is $\| x \|^2 = N$ times the coefficient of projection onto x of y *advanced* by n samples (shifted circularly to the *left* by n samples). The time shift n is called the *correlation lag*, and $\overline{x(m)}y(m+n)$ is called a *lagged product*. Applications of correlation are discussed in §8.4.

7.2.6 Stretch Operator

Unlike all previous operators, the $\text{STRETCH}_L()$ operator maps a length N signal to a length $M \triangleq LN$ signal, where L and N are integers. We use "m" instead of "n" as the time index to underscore this fact.

A *stretch by factor L* is defined by

$$\text{STRETCH}_{L,m}(x) \triangleq \begin{cases} x(m/L), & m/L \text{ an integer} \\ 0, & m/L \text{ non-integer.} \end{cases}$$

Thus, to stretch a signal by the factor L, insert $L-1$ zeros between each pair of samples. An example of a stretch by factor three is shown in Fig. 7.6. The example is

$$\text{STRETCH}_3([4,1,2]) = [4,0,0,1,0,0,2,0,0].$$

The stretch operator is used to describe and analyze *upsampling*, that is, increasing the sampling rate by an integer factor. A stretch by K followed by lowpass filtering to the frequency band $\omega \in (-\pi/K, \pi/K)$ implements *ideal bandlimited interpolation* (introduced in Appendix D).

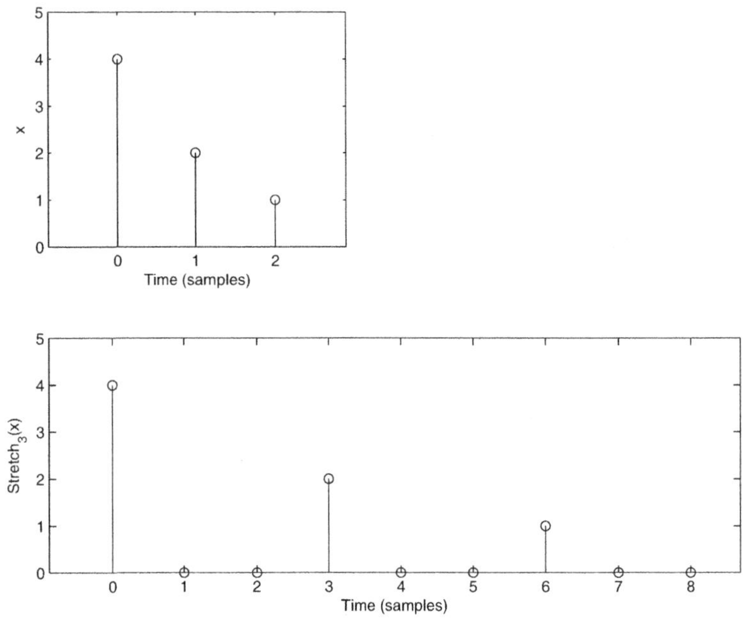

Figure 7.6: Illustration of STRETCH$_3(x)$.

7.2.7 Zero Padding

Zero padding consists of extending a signal (or spectrum) with zeros. It maps a length N signal to a length $M > N$ signal, but N need not divide M.

Definition:

$$\text{ZeroPad}_{M,m}(x) \triangleq \begin{cases} x(m), & |m| < N/2 \\ 0, & \text{otherwise} \end{cases} \tag{7.4}$$

where $m = 0, \pm 1, \pm 2, \dots, \pm M_h$, with $M_h \triangleq (M-1)/2$ for M odd, and $M/2 - 1$ for M even. For example,

$$\text{ZeroPad}_{10}([1,2,3,4,5]) = [1,2,3,0,0,0,0,0,4,5].$$

In this example, the first sample corresponds to time 0, and five zeros have been inserted between the samples corresponding to times $n = 2$ and $n = -2$.

Figure 7.7 illustrates zero padding from length $N = 5$ out to length $M = 11$. Note that x and n could be replaced by X and k in the figure caption.

Note that we have unified the time-domain and frequency-domain definitions of zero-padding by interpreting the original time axis $[0, 1, \dots, N-1]$ as indexing positive-time samples from 0 to $N/2-1$ (for N even), and negative times in the interval $n \in [N - N/2 + 1, N - 1] \equiv [-N/2 + 1, -1]$.[8] Furthermore, we require $x(N/2) \equiv x(-N/2) = 0$ when N is even, while odd N requires no such restriction. In practice, we often prefer to interpret time-domain samples as extending from 0 to $N - 1$, *i.e.*, with no negative-time samples. For this case, we define "causal zero padding" as described below.

Causal (Periodic) Signals

A signal $x \in \mathbf{C}^N$ may be defined as *causal* when $x(n) = 0$ for all "negative-time" samples (*e.g.*, for $n = -1, -2, \dots, -N/2$ when N is even). Thus, the signal $x = [1,2,3,0,0] \in \mathbf{R}^5$ is causal while $x = [1,2,3,4,0]$ is not. For causal signals, zero-padding is equivalent to simply *appending* zeros to the original signal. For example,

$$\text{ZeroPad}_{10}([1,2,3,0,0]) = [1,2,3,0,0,0,0,0,0,0].$$

Therefore, when we simply append zeros to the end of signal, we call it *causal zero padding*.

---

[8] The symbol "$\equiv$" means "is equivalent to".

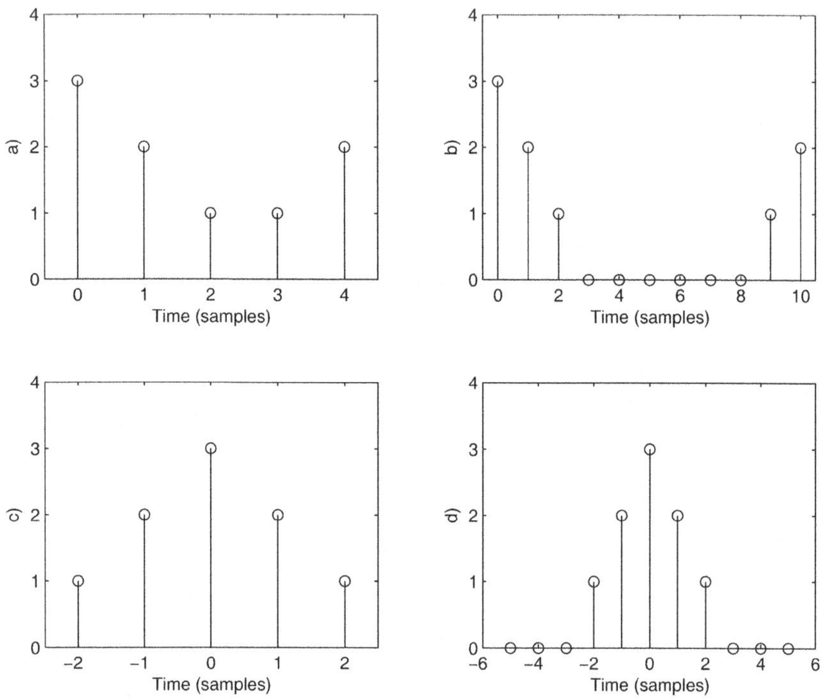

Figure 7.7: Illustration of zero padding: a) Original signal (or spectrum) $x = [3, 2, 1, 1, 2]$ plotted over the domain $n \in [0, N-1]$ where $N = 5$ (*i.e.*, as the samples would normally be held in a computer array). b) $\text{ZEROPAD}_{11}(x)$. c) The same signal x plotted over the domain $n \in [-(N-1)/2, (N-1)/2]$ which is more natural for interpreting negative times (frequencies). d) $\text{ZEROPAD}_{11}(x)$ plotted over the zero-centered domain.

7.2.8 Causal Zero Padding

In practice, a signal $x \in \mathbf{R}^N$ is often an N-sample *frame* of data taken from some longer signal, and its true starting time can be anything. In such cases, it is common to treat the start-time of the frame as zero, with no negative-time samples. In other words, $x(n)$ represents an N-sample signal-segment that is translated in time to start at time 0. In this case (no negative-time samples in the frame), it is proper to zero-pad by simply appending zeros at the end of the frame. Thus, we define *e.g.*,

$$\text{CausalZeroPad}_{10}([1,2,3,4,5]) = [1,2,3,4,5,0,0,0,0,0].$$

Causal zero-padding should not be used on a spectrum of a real signal because, as we will see in §7.4.3 below, the magnitude spectrum of every real signal is symmetric about frequency zero. For the same reason, we cannot simply append zeros in the time domain when the signal frame is considered to include negative-time samples, as in "zero-centered FFT processing" (discussed in Book IV [67]). Nevertheless, in practice, appending zeros is perhaps the most common form of zero-padding. It is implemented automatically, for example, by the matlab function `fft(x,N)` when the FFT size `N` exceeds the length of the signal vector `x`.

In summary, we have defined two types of zero-padding that arise in practice, which we may term "causal" and "zero-centered" (or "zero-phase", or even "periodic"). The zero-centered case is the more natural with respect to the mathematics of the DFT, so it is taken as the "official" definition of ZeroPad(). In both cases, however, when properly used, we will have the basic Fourier theorem (§7.4.12 below) stating that *zero-padding in the time domain corresponds to ideal bandlimited interpolation in the frequency domain*, and vice versa.

7.2.9 Zero Padding Applications

Zero padding in the time domain is used extensively in practice to compute heavily *interpolated* spectra by taking the DFT of the zero-padded signal. Such spectral interpolation is ideal when the original signal is *time limited* (nonzero only over some finite duration spanned by the orignal samples).

Note that the *time-limited* assumption directly contradicts our usual assumption of *periodic extension*. As mentioned in §6.7 on page 105, the interpolation of a periodic signal's spectrum from its harmonics is always zero; that is, there is no spectral energy, in principle, between the harmonics of a periodic signal, and a periodic signal cannot be time-limited unless it is the zero signal. On the other hand, the interpolation of a time-limited signal's spectrum is nonzero almost everywhere between the original spectral samples. Thus, zero-padding is often used when analyzing data from a non-periodic signal in blocks, and each block, or *frame*, is treated as a finite-duration signal which can be zero-padded on either side with any number of zeros. In summary, the use of zero-padding corresponds to the *time-limited assumption* for the data frame, and more zero-padding yields denser interpolation of the frequency samples around the unit circle.

Sometimes people will say that zero-padding in the time domain yields higher spectral *resolution* in the frequency domain. However, signal processing practitioners should not say that, because "resolution" in signal processing refers to the ability to "resolve" closely spaced features in a spectrum analysis (see Book IV [67] for details). The usual way to increase spectral resolution is to take a longer DFT *without* zero padding—*i.e.*, look at more data. In the field of *graphics*, the term resolution refers to pixel density, so the common terminology confusion is reasonable. However, remember that in signal processing, zero-padding in one domain corresponds to a higher interpolation-density in the other domain—not a higher resolution.

7.2.10 Ideal Spectral Interpolation

Using Fourier theorems, we will be able to show (§7.4.12) that *zero padding in the time domain* gives *exact bandlimited interpolation in the frequency domain.*[9] In other words, for truly time-limited signals x, taking the DFT of the entire nonzero portion of x extended by zeros yields *exact interpolation* of the complex spectrum—not an approximation (ignoring computational round-off error in the DFT itself). Because the fast Fourier transform (FFT) is so efficient, zero-padding followed by an FFT is a highly practical method for interpolating spectra of finite-duration signals, and is used extensively in practice.

Before we can interpolate a spectrum, we must be clear on what a "spectrum" really is. As discussed in Chapter 6, the *spectrum* of a signal $x(\cdot)$ at frequency ω is defined as a complex number $X(\omega)$ computed using the inner product

$$X(\omega) \triangleq \langle x, s_\omega \rangle \triangleq \sum_{\text{all } n} x(n)e^{-j\omega nT}.$$

That is, $X(\omega)$ is the unnormalized coefficient of projection of x onto the sinusoid s_ω at frequency ω. When $\omega = \omega_k = 2\pi f_s k/N$, for $k = 0, 1, \ldots, N-1$, we obtain the special set of spectral samples known as the DFT. For other

[9]Similarly, zero padding in the frequency domain gives what we may call "periodic interpolation" in the time domain which is exact in the DFT case only for periodic signals having a time-domain period equal to the DFT length. (See §6.7 on page 105.)

values of ω, we obtain spectral points in between the DFT samples. Interpolating DFT samples should give the same result. It is straightforward to show that this ideal form of interpolation is what we call *bandlimited interpolation*, as discussed further in Appendix D and in Book IV [67] of this series.

7.2.11 Interpolation Operator

The *interpolation operator* $\text{INTERP}_L()$ interpolates a signal by an integer factor L using bandlimited interpolation. For frequency-domain signals $X(\omega_k)$, $k = 0, 1, 2, \ldots, N-1$, we may write spectral interpolation as follows:

$$\begin{aligned} \text{INTERP}_{L,k'}(X) &\triangleq X(\omega_{k'}), \text{ where} \\ \omega_{k'} &= 2\pi k'/M, \; k' = 0, 1, 2, \ldots, M-1, \\ M &\triangleq LN. \end{aligned}$$

Since $X(\omega_k) \triangleq \text{DFT}_{N,k}(x)$ is initially only defined over the N roots of unity in the z plane, while $X(\omega_{k'})$ is defined over $M = LN$ roots of unity, we define $X(\omega_{k'})$ for $\omega_{k'} \neq \omega_k$ by ideal bandlimited interpolation (specifically *time*-limited spectral interpolation in this case).

For time-domain signals $x(n)$, exact interpolation is similarly bandlimited interpolation, as derived in Appendix D.

7.2.12 Repeat Operator

Like the $\text{STRETCH}_L()$ and $\text{INTERP}_L()$ operators, the $\text{REPEAT}_L()$ operator maps a length N signal to a length $M \triangleq LN$ signal:

Definition: The *repeat L times* operator is defined for any $x \in \mathbf{C}^N$ by

$$\text{REPEAT}_{L,m}(x) \triangleq x(m), \qquad m = 0, 1, 2, \ldots, M-1,$$

where $M \triangleq LN$, and indexing of x is modulo N (periodic extension). Thus, the $\text{REPEAT}_L()$ operator simply repeats its input signal L times.[10] An example of $\text{REPEAT}_2(x)$ is shown in Fig. 7.8. The example is

$$\text{REPEAT}_2([0, 2, 1, 4, 3, 1]) = [0, 2, 1, 4, 3, 1, 0, 2, 1, 4, 3, 1].$$

[10] You might wonder why we need this since all indexing in $\mathbf{C}^N$ is defined modulo N already. The answer is that $\text{REPEAT}_L()$ formally expresses a mapping from the space $\mathbf{C}^N$ of length N signals to the space $\mathbf{C}^M$ of length $M = LN$ signals. The $\text{REPEAT}_L()$ operator could alternatively be defined as the *scaling operator* $\text{SCALE}_{L,n}(x) = x(n)$, $n = 0, 1, 2, \ldots, NL-1$, where L is any positive integer (note the increase in signal length from N to NL samples). Such a definition would better suggest the related *continuous-time* Fourier theorems regarding time/frequency scaling (see Appendix C, especially §C.2 on page 214).

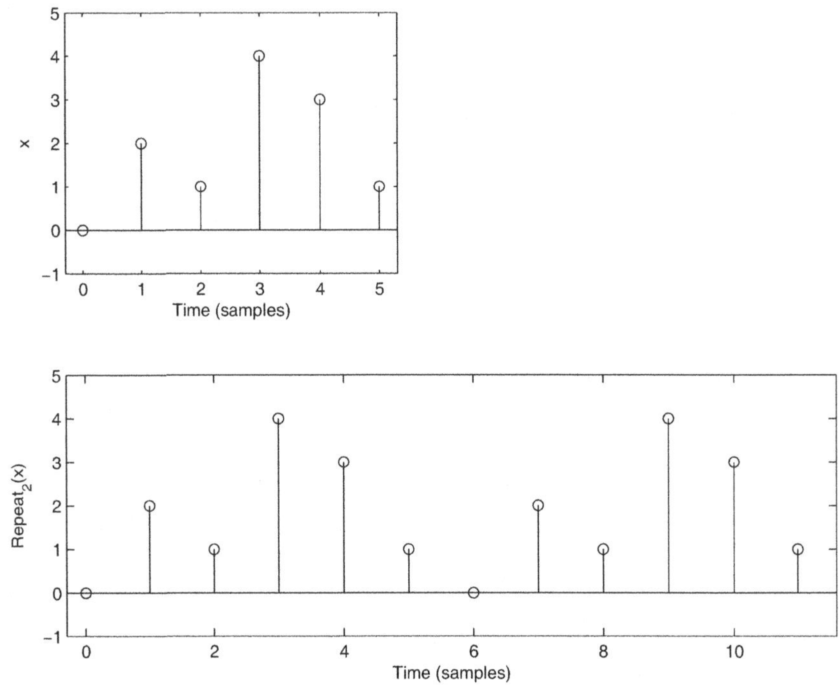

Figure 7.8: Illustration of $\text{Repeat}_2(x)$.

A frequency-domain example is shown in Fig. 7.9. Figure 7.9a shows the original spectrum X, Fig. 7.9b shows the same spectrum plotted over the unit circle in the z plane, and Fig. 7.9c shows $\text{Repeat}_3(X)$. The $z = 1$ point (dc) is on the right-rear face of the enclosing box. Note that when viewed as centered about $k = 0$, X is a somewhat "triangularly shaped" spectrum. We see three copies of this shape in $\text{Repeat}_3(X)$.

The repeat operator is used to state the Fourier theorem

$$\text{Stretch}_L \longleftrightarrow \text{Repeat}_L,$$

where Stretch_L is defined in §7.2.6 on page 129. That is, when you stretch a signal by the factor L (inserting zeros between the original samples), its spectrum is repeated L times around the unit circle. The simple proof is given on page 157.

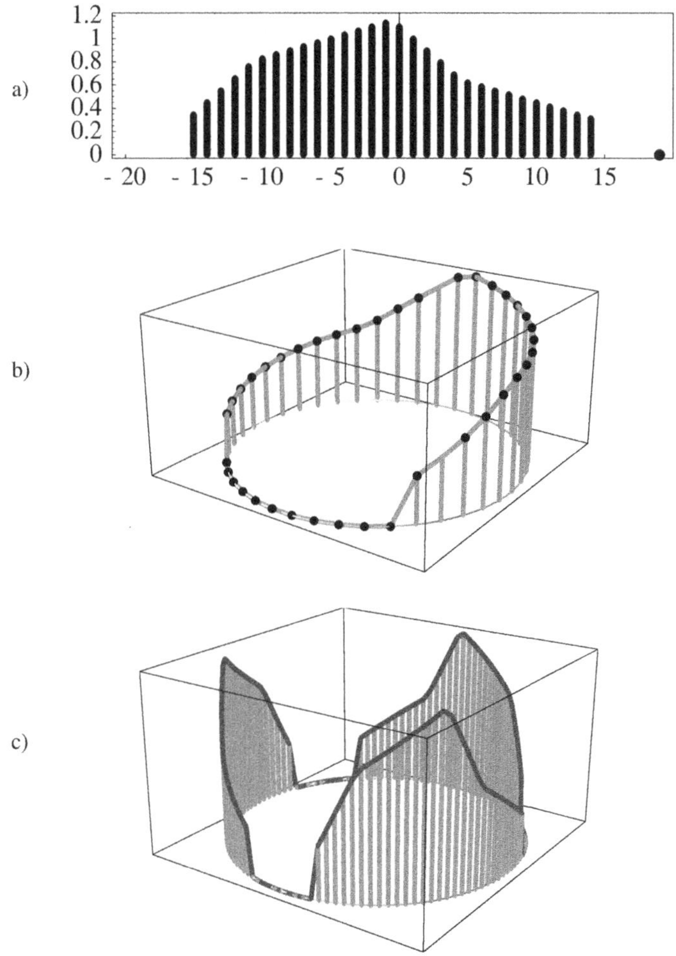

Figure 7.9: Illustration of $\text{REPEAT}_3(X)$. a) Conventional plot of X. b) Plot of X over the unit circle in the z plane. c) $\text{REPEAT}_3(X)$.

7.2.13 Downsampling Operator

Downsampling by L (also called *decimation* by L) is defined for $x \in \mathbf{C}^N$ as taking every Lth sample, starting with sample zero:

$$\begin{aligned} \text{Downsample}_{L,m}(x) &\triangleq x(mL), \\ m &= 0, 1, 2, \ldots, M-1 \\ N &= LM. \end{aligned}$$

The $\text{Downsample}_L()$ operator maps a length $N = LM$ signal down to a length M signal. It is the inverse of the $\text{Stretch}_L()$ operator (but not vice versa), *i.e.*,

$$\begin{aligned} \text{Downsample}_L(\text{Stretch}_L(x)) &= x \\ \text{Stretch}_L(\text{Downsample}_L(x)) &\neq x \quad \text{(in general)}. \end{aligned}$$

The stretch and downsampling operations do not commute because they are linear *time-varying* operators. They can be modeled using time-varying *switches* controlled by the sample index n.

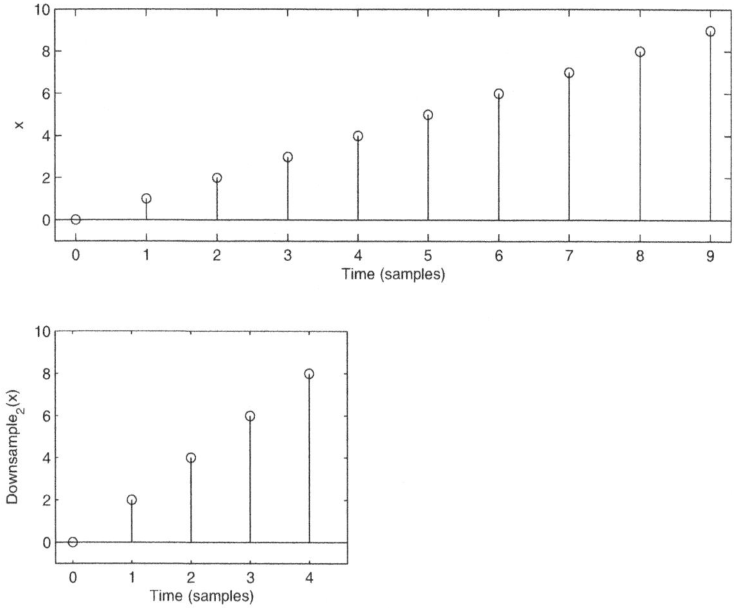

Figure 7.10: Illustration of $\text{Downsample}_2(x)$.

The following example of $\text{Downsample}_2(x)$ is illustrated in Fig. 7.10:

$$\text{Downsample}_2([0, 1, 2, 3, 4, 5, 6, 7, 8, 9]) = [0, 2, 4, 6, 8].$$

Note that the term "downsampling" may also refer to the more elaborate process of *sampling-rate conversion* to a lower sampling rate, in which a signal's sampling rate is lowered by resampling using bandlimited interpolation. To distinguish these cases, we can call this *bandlimited downsampling*, because a lowpass-filter is needed, in general, prior to downsampling so that *aliasing* is avoided. This topic is address in Appendix D. Early sampling-rate converters were in fact implemented using the STRETCH_L operation, followed by an appropriate lowpass filter, followed by DOWNSAMPLE_M, in order to implement a sampling-rate conversion by the factor L/M.

7.2.14 Alias Operator

Aliasing occurs when a signal is *undersampled.* If the signal sampling rate f_s is too low, we get *frequency-domain aliasing.*

The topic of aliasing normally arises in the context of *sampling* a continuous-time signal. The *sampling theorem* (Appendix D) says that we will have no aliasing due to sampling as long as the sampling rate is higher than twice the highest frequency present in the signal being sampled.

In this chapter, we are considering only discrete-time signals, in order to keep the math as simple as possible. Aliasing in this context occurs when a discrete-time signal is downsampled to reduce its sampling rate. You can think of continuous-time sampling as the limiting case for which the starting sampling rate is infinity.

An example of aliasing is shown in Fig. 7.11. In the figure, the high-frequency sinusoid is indistinguishable from the lower-frequency sinusoid due to aliasing. We say the higher frequency *aliases* to the lower frequency.

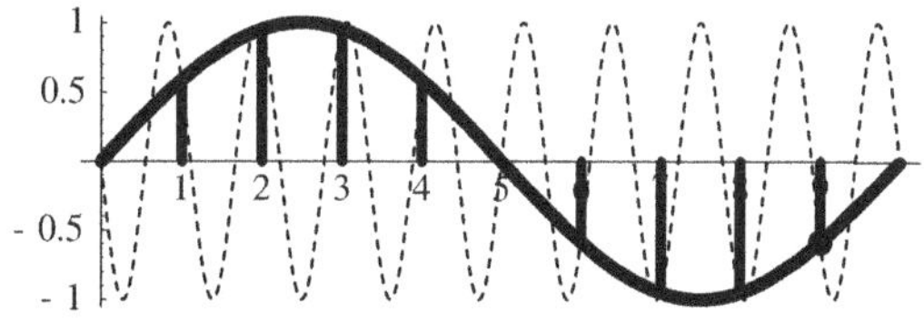

Figure 7.11: Example of frequency-domain aliasing due to undersampling in the time domain.

Undersampling in the frequency domain gives rise to *time-domain aliasing*. If time or frequency is not specified, the term "aliasing" normally means frequency-domain aliasing (due to undersampling in the time domain).

The *aliasing operator* for N-sample signals $x \in \mathbf{C}^N$ is defined by

$$\begin{aligned}
\text{ALIAS}_{L,m}(x) &\triangleq \sum_{l=0}^{L-1} x\,(m+lM)\,, \\
m &= 0,1,2,\ldots,M-1, \\
N &= LM.
\end{aligned}$$

Like the $\text{DOWNSAMPLE}_L()$ operator, the $\text{ALIAS}_L()$ operator maps a length $N = LM$ signal down to a length M signal. A way to think of it is to partition the original N samples into L blocks of length M, with the first block extending from sample 0 to sample $M-1$, the second block from M to $2M-1$, etc. Then just add up the blocks. This process is called *aliasing*. If the original signal x is a time signal, it is called *time-domain aliasing*; if it is a spectrum, we call it *frequency-domain aliasing*, or just aliasing. Note that aliasing is *not invertible* in general. Once the blocks are added together, it is usually not possible to recover the original blocks.

Example:

$$\begin{aligned}
\text{ALIAS}_2([0,1,2,3,4,5]) &= [0,1,2]+[3,4,5]=[3,5,7] \\
\text{ALIAS}_3([0,1,2,3,4,5]) &= [0,1]+[2,3]+[4,5]=[6,9]
\end{aligned}$$

The alias operator is used to state the Fourier theorem (§7.4.11)

$$\text{DOWNSAMPLE}_L \longleftrightarrow \frac{1}{L}\text{ALIAS}_L.$$

That is, when you downsample a signal by the factor L, its spectrum is aliased by the factor L.

Figure 7.12 shows the result of ALIAS_2 applied to $\text{REPEAT}_3(X)$ from Figure 7.9c. Imagine the spectrum of Fig. 7.12a as being plotted on a piece of paper rolled to form a cylinder, with the edges of the paper meeting at $z = 1$ (upper right corner of Fig. 7.12a). Then the ALIAS_2 operation can be simulated by rerolling the cylinder of paper to cut its circumference in half. That is, reroll it so that at every point, *two* sheets of paper are in contact at all points on the new, narrower cylinder. Now, simply add the values on the two overlapping sheets together, and you have the ALIAS_2 of the original spectrum on the unit circle. To alias by 3, we would shrink the cylinder further until the paper edges again line up, giving three layers of paper in the cylinder, and so on.

Figure 7.12b shows what is plotted on the first circular wrap of the cylinder of paper, and Fig. 7.12c shows what is on the second wrap. These are overlaid in Fig. 7.12d and added together in Fig. 7.12e. Finally, Figure 7.12f shows both the addition and the overlay of the two components.

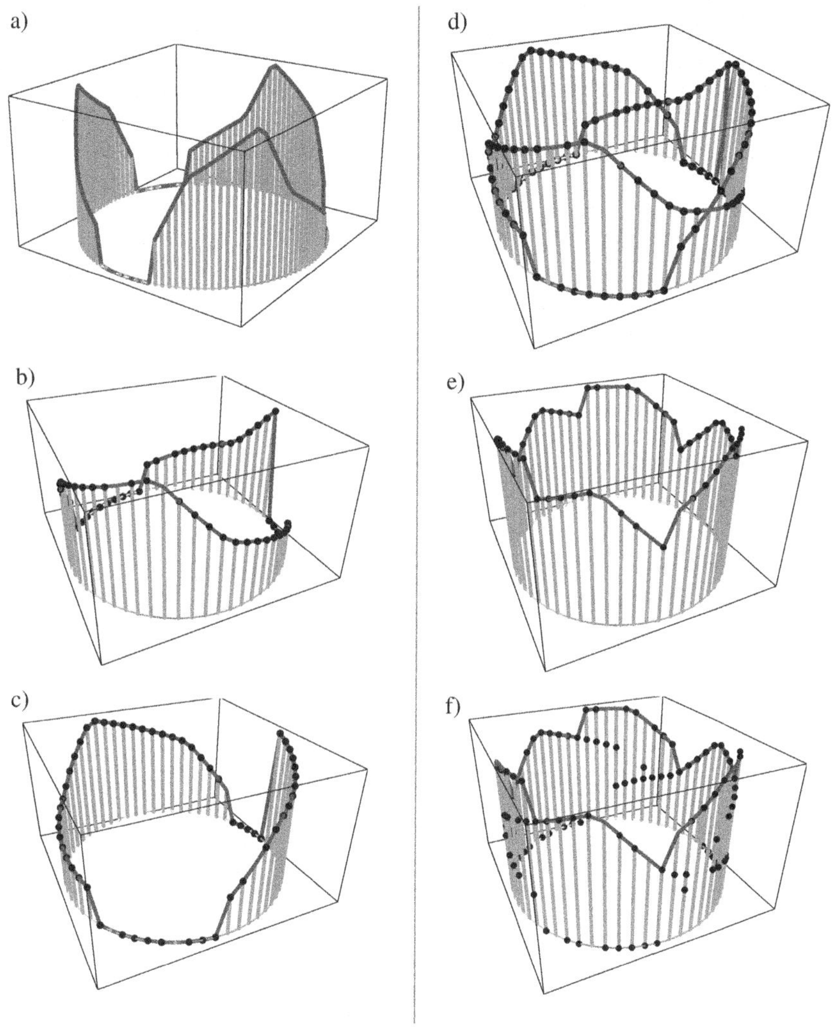

Figure 7.12: Illustration of aliasing in the frequency domain. a) $\text{REPEAT}_3(X)$ from Fig. 7.9c. b) First half of the original unit circle (0 to π) wrapped around the new, smaller unit circle (which is magnified to the original size). c) Second half (π to 2π), also wrapped around the new unit circle. d) Overlay of components to be summed. e) Sum of components (the aliased spectrum). f) Both sum and overlay.

We say that the second component (Fig. 7.12c) "aliases" to new frequency components, while the first component (Fig. 7.12b) is considered to be at its original frequencies. If the unit circle of Fig. 7.12a covers frequencies 0 to f_s, all other unit circles (Fig. 7.12b-c) cover frequencies 0 to $f_s/2$.

In general, aliasing by the factor K corresponds to a *sampling-rate reduction* by the factor K. To prevent aliasing when reducing the sampling rate, an *anti-aliasing lowpass filter* is generally used. The lowpass filter attenuates all signal components at frequencies outside the interval $[-f_s/(2K), f_s/(2K)]$ so that all frequency components which would alias are first removed.

Conceptually, in the frequency domain, the unit circle is reduced by ALIAS_2 to a unit circle *half* the original size, where the two halves are summed. The inverse of aliasing is then "repeating" which should be understood as *increasing* the unit circle circumference using "periodic extension" to generate "more spectrum" for the larger unit circle. In the time domain, on the other hand, downsampling is the inverse of the stretch operator. We may interchange "time" and "frequency" and repeat these remarks. All of these relationships are precise only for *integer* stretch/downsampling/-aliasing/repeat factors; in continuous time and frequency, the restriction to integer factors is removed, and we obtain the (simpler) *scaling theorem* (proved in §C.2 on page 214).

7.3 Even and Odd Functions

Some of the Fourier theorems can be succinctly expressed in terms of even and odd symmetries.

Definition: A function $f(n)$ is said to be *even* if $f(-n) = f(n)$.

An even function is also *symmetric*, but the term symmetric applies also to functions symmetric about a point other than 0.

Definition: A function $f(n)$ is said to be *odd* if $f(-n) = -f(n)$.

An odd function is also called *antisymmetric*.

Note that every finite odd function $f(n)$ must satisfy $f(0) = 0$.[11] Moreover, for any $x \in \mathbf{C}^N$ with N even, we also have $x(N/2) = 0$ since $x(N/2) = -x(-N/2) = -x(-N/2 + N) = -x(N/2)$; that is, $N/2$ and $-N/2$ index the same point when N is even.

[11] The function $f(x) = 1/x$ is also considered odd, ignoring the singularity at $x = 0$.

Theorem: Every function $f(n)$ can be decomposed into a sum of its even part $f_e(n)$ and odd part $f_o(n)$, where

$$\begin{aligned} f_e(n) &\triangleq \frac{f(n)+f(-n)}{2} \\ f_o(n) &\triangleq \frac{f(n)-f(-n)}{2}. \end{aligned}$$

Proof: In the above definitions, f_e is even and f_o is odd by construction. Summing, we have

$$f_e(n)+f_o(n) = \frac{f(n)+f(-n)}{2} + \frac{f(n)-f(-n)}{2} = f(n).$$

Theorem: The product of even functions is even, the product of odd functions is even, and the product of an even times an odd function is odd.

Proof: Readily shown.

Since even times even is even, odd times odd is even, and even times odd is odd, we can think of even as (+) and odd as (−):

$$\begin{aligned} (+)\cdot(+) &= (+) \\ (-)\cdot(-) &= (+) \\ (+)\cdot(-) &= (-) \\ (-)\cdot(+) &= (-) \end{aligned}$$

Example: $\cos(\omega_k n)$, $n \in \mathbf{Z}$, is an *even* signal since $\cos(-\theta) = \cos(\theta)$.

Example: $\sin(\omega_k n)$ is an *odd* signal since $\sin(-\theta) = -\sin(\theta)$.

Example: $\cos(\omega_k n)\cdot\sin(\omega_l n)$ is an *odd* signal (even times odd).

Example: $\sin(\omega_k n)\cdot\sin(\omega_l n)$ is an *even* signal (odd times odd).

Theorem: The sum of all the samples of an odd signal x_o in $\mathbf{C}^N$ is zero.

Proof: This is readily shown by writing the sum as $x_o(0)+[x_o(1)+x_o(-1)]+\cdots+x(N/2)$, where the last term only occurs when N is even. Each term so written is zero for an odd signal x_o.

Example: For all DFT sinusoidal frequencies $\omega_k = 2\pi k/N$,

$$\sum_{n=0}^{N-1} \sin(\omega_k n)\cos(\omega_k n) = 0,\ k=0,1,2,\ldots,N-1.$$

More generally,

$$\sum_{n=0}^{N-1} x_e(n)x_o(n) = 0,$$

for *any* even signal x_e and odd signal x_o in $\mathbf{C}^N$. In terms of inner products (§5.9 on page 80), we may say that the even part of every real signal is orthogonal to its odd part:

$$\langle x_e, x_o \rangle = 0$$

7.4 Fourier Theorems

In this section the main Fourier theorems are stated and proved. It is no small matter how simple these theorems are in the DFT case relative to the other three cases (DTFT, Fourier transform, and Fourier series, as defined in Appendix B). When infinite summations or integrals are involved, the conditions for the existence of the Fourier transform can be quite difficult to characterize mathematically. Mathematicians have expended a considerable effort on such questions. By focusing primarily on the DFT case, we are able to study the essential concepts conveyed by the Fourier theorems without getting involved with mathematical difficulties.

7.4.1 Linearity

Theorem: For any $x, y \in \mathbf{C}^N$ and $\alpha, \beta \in \mathbf{C}$, the DFT satisfies

$$\boxed{\alpha x + \beta y \longleftrightarrow \alpha X + \beta Y}$$

where $X \triangleq \text{DFT}(x)$ and $Y \triangleq \text{DFT}(y)$, as always in this book. Thus, the DFT is a *linear operator*.

Proof:

$$\begin{aligned}
\text{DFT}_k(\alpha x + \beta y) &\triangleq \sum_{n=0}^{N-1} [\alpha x(n) + \beta y(n)] e^{-j2\pi nk/N} \\
&= \sum_{n=0}^{N-1} \alpha x(n) e^{-j2\pi nk/N} + \sum_{n=0}^{N-1} \beta y(n) e^{-j2\pi nk/N} \\
&= \alpha \sum_{n=0}^{N-1} x(n) e^{-j2\pi nk/N} + \beta \sum_{n=0}^{N-1} y(n) e^{-j2\pi nk/N} \\
&\triangleq \alpha X + \beta Y
\end{aligned}$$

7.4.2 Conjugation and Reversal

Theorem: For any $x \in \mathbf{C}^N$,

$$\boxed{\overline{x} \longleftrightarrow \text{FLIP}(\overline{X}).}$$

Proof:

$$\begin{aligned}
\text{DFT}_k(\overline{x}) &\triangleq \sum_{n=0}^{N-1} \overline{x(n)} e^{-j2\pi nk/N} = \overline{\sum_{n=0}^{N-1} x(n) e^{j2\pi nk/N}} \\
&= \overline{\sum_{n=0}^{N-1} x(n) e^{-j2\pi n(-k)/N}} \triangleq \text{FLIP}_k(\overline{X})
\end{aligned}$$

Theorem: For any $x \in \mathbf{C}^N$,

$$\boxed{\text{FLIP}(\overline{x}) \longleftrightarrow \overline{X}.}$$

Proof: Making the change of summation variable $m \triangleq N - n$, we get

$$\begin{aligned}
\text{DFT}_k(\text{FLIP}(\overline{x})) &\triangleq \sum_{n=0}^{N-1} \overline{x(N-n)} e^{-j2\pi nk/N} = \sum_{m=N}^{1} \overline{x(m)} e^{-j2\pi(N-m)k/N} \\
&= \sum_{m=0}^{N-1} \overline{x(m)} e^{j2\pi mk/N} = \overline{\sum_{m=0}^{N-1} x(m) e^{-j2\pi mk/N}} \triangleq \overline{X(k)}.
\end{aligned}$$

Theorem: For any $x \in \mathbf{C}^N$,

$$\boxed{\text{FLIP}(x) \longleftrightarrow \text{FLIP}(X).}$$

Proof:

$$\begin{aligned}
\text{DFT}_k[\text{FLIP}(x)] &\triangleq \sum_{n=0}^{N-1} x(N-n) e^{-j2\pi nk/N} = \sum_{m=0}^{N-1} x(m) e^{-j2\pi(N-m)k/N} \\
&= \sum_{m=0}^{N-1} x(m) e^{j2\pi mk/N} \triangleq X(-k) \triangleq \text{FLIP}_k(X)
\end{aligned}$$

Corollary: For any $x \in \mathbf{R}^N$,

$$\boxed{\text{FLIP}(x) \longleftrightarrow \overline{X}} \quad (x \text{ real}).$$

Proof: Picking up the previous proof at the third formula, remembering that x is real,

$$\sum_{m=0}^{N-1} x(m)e^{j2\pi mk/N} = \overline{\sum_{m=0}^{N-1} \overline{x(m)}e^{-j2\pi mk/N}} = \overline{\sum_{m=0}^{N-1} x(m)e^{-j2\pi mk/N}} \triangleq \overline{X(k)}$$

when $x(m)$ is real.

Thus, *conjugation* in the frequency domain corresponds to *reversal* in the time domain. Another way to say it is that *negating spectral phase flips the signal around backwards in time.*

Corollary: For any $x \in \mathbf{R}^N$,

$$\boxed{\text{FLIP}(X) = \overline{X}} \quad (x \text{ real}).$$

Proof: This follows from the previous two cases.

Definition: The property $X(-k) = \overline{X(k)}$ is called *Hermitian symmetry* or "conjugate symmetry." If $X(-k) = -\overline{X(k)}$, it may be called *skew-Hermitian*.

Another way to state the preceding corollary is

$$\boxed{x \in \mathbf{R}^N \longleftrightarrow X \text{ is Hermitian.}}$$

7.4.3 Symmetry

In the previous section, we found $\text{FLIP}(X) = \overline{X}$ when x is real. This fact is of high practical importance. It says that the spectrum of every real signal is *Hermitian*. Due to this symmetry, we may discard all negative-frequency spectral samples of a real signal and regenerate them later if needed from the positive-frequency samples. Also, spectral plots of real signals are normally displayed only for positive frequencies; *e.g.*, spectra of sampled signals are normally plotted over the range 0 Hz to $f_s/2$ Hz. On the other hand, the spectrum of a *complex* signal must be shown, in general, from $-f_s/2$ to $f_s/2$ (or from 0 to f_s), since the positive and negative frequency components of a complex signal are independent.

Recall from §7.3 on page 142 that a signal $x(n)$ is said to be *even* if $x(-n) = x(n)$, and *odd* if $x(-n) = -x(n)$. Below are are Fourier theorems pertaining to even and odd signals and/or spectra.

Theorem: If $x \in \mathbf{R}^N$, then $\text{re}\{X\}$ is *even* and $\text{im}\{X\}$ is *odd*.

Proof: This follows immediately from the conjugate symmetry of X for real signals x.

Theorem: If $x \in \mathbf{R}^N$, $|X|$ is *even* and $\angle X$ is *odd*.

Proof: This follows immediately from the conjugate symmetry of X expressed in polar form $X(k) = |X(k)|\, e^{j\angle X(k)}$.

The conjugate symmetry of spectra of real signals is perhaps the most important symmetry theorem. However, there are a couple more we can readily show:

Theorem: An even signal has an even transform:

$$\boxed{x \text{ even} \longleftrightarrow X \text{ even}}$$

Proof: Express x in terms of its real and imaginary parts by $x \triangleq x_r + jx_i$. Note that for a complex signal x to be even, both its real and imaginary parts must be even. Then

$$\begin{aligned}
X(k) &\triangleq \sum_{n=0}^{N-1} x(n)e^{-j\omega_k n} \\
&= \sum_{n=0}^{N-1} [x_r(n) + jx_i(n)]\cos(\omega_k n) - j[x_r(n) + jx_i(n)]\sin(\omega_k n) \\
&= \sum_{n=0}^{N-1} [x_r(n)\cos(\omega_k n) + x_i(n)\sin(\omega_k n)] \\
&\quad + j[x_i(n)\cos(\omega_k n) - x_r(n)\sin(\omega_k n)]. \qquad (7.5)
\end{aligned}$$

Let even_n denote a function that is even in n, such as $f(n) = n^2$, and let odd_n denote a function that is odd in n, such as $f(n) = n^3$, Similarly, let even_{nk} denote a function of n and k that is even in both n and k, such as $f(n,k) = n^2k^2$, and odd_{nk} mean odd in both n and k. Then appropriately labeling each term in the last formula above gives

$$\begin{aligned}
X(k) &= \sum_{n=0}^{N-1} \text{even}_n \cdot \text{even}_{nk} + \underbrace{\sum_{n=0}^{N-1} \text{even}_n \cdot \text{odd}_{nk}}_{\text{sums to } 0} \\
&+ j \cdot \sum_{n=0}^{N-1} \text{even}_n \cdot \text{even}_{nk} - \underbrace{j\sum_{n=0}^{N-1} \text{even}_n \cdot \text{odd}_{nk}}_{\text{sums to } 0} \\
&= \sum_{n=0}^{N-1} \text{even}_n \cdot \text{even}_{nk} + j \cdot \sum_{n=0}^{N-1} \text{even}_n \cdot \text{even}_{nk} \\
&= \text{even}_k + j \cdot \text{even}_k = \text{even}_k.
\end{aligned}$$

Theorem: A real even signal has a real even transform:

$$\boxed{x \text{ real and even} \longleftrightarrow X \text{ real and even}} \tag{7.6}$$

Proof: This follows immediately from setting $x_i(n) = 0$ in the preceding proof. From Eq. (7.5), we are left with

$$X(k) = \sum_{n=0}^{N-1} x_r(n)\cos(\omega_k n).$$

Thus, the DFT of a real and even function reduces to a type of *cosine transform*,[12]

Instead of adapting the previous proof, we can show it directly:

$$\begin{aligned}
X(k) &\triangleq \sum_{n=0}^{N-1} x(n)e^{-j\omega_k n} = \sum_{n=0}^{N-1} x(n)\cos(\omega_k n) + j\sum_{n=0}^{N-1} x(n)\sin(\omega_k n) \\
&= \sum_{n=0}^{N-1} x(n)\cos(\omega_k n) \qquad (\textstyle\sum_n \text{even}_n \cdot \text{odd}_n = \sum \text{odd}_n = 0) \\
&= \sum_{n=0}^{N-1} \text{even}_n \cdot \text{even}_{nk} = \sum_{n=0}^{N-1} \text{even}_{nk} = \text{even}_k
\end{aligned}$$

Definition: A signal with a real spectrum (such as any real, even signal) is often called a *zero phase signal*. However, note that when the spectrum goes *negative* (which it can), the phase is really $\pm\pi$, not 0. When a real spectrum is positive at dc (*i.e.*, $X(0) > 0$), it is then truly zero-phase over at least some band containing dc (up to the first zero-crossing in frequency). When the phase switches between 0 and π at the zero-crossings of the (real) spectrum, the spectrum oscillates between being zero phase and "constant phase". We can say that all real spectra are *piecewise constant-phase spectra*, where the two constant values are 0 and π (or $-\pi$, which is the same phase as $+\pi$). In practice, such zero-crossings typically occur at low magnitude, such as in the "side-lobes" of the DTFT of a "zero-centered symmetric window" used for spectrum analysis (see Chapter 8 and Book IV [67]).

[12]The discrete cosine transform (DCT) used most often in applications is defined somewhat differently (see §A.6.1 on page 203), but this is one variant.

7.4.4 Shift Theorem

Theorem: For any $x \in \mathbf{C}^N$ and any integer Δ,

$$\boxed{\text{DFT}_k[\text{SHIFT}_\Delta(x)] = e^{-j\omega_k\Delta}X(k).}$$

Proof:

$$\begin{aligned}
\text{DFT}_k[\text{SHIFT}_\Delta(x)] &\triangleq \sum_{n=0}^{N-1} x(n-\Delta)e^{-j2\pi nk/N} \\
&= \sum_{m=-\Delta}^{N-1-\Delta} x(m)e^{-j2\pi(m+\Delta)k/N} \qquad (m \triangleq n-\Delta) \\
&= \sum_{m=0}^{N-1} x(m)e^{-j2\pi mk/N}e^{-j2\pi k\Delta/N} \\
&= e^{-j2\pi\Delta k/N}\sum_{m=0}^{N-1} x(m)e^{-j2\pi mk/N} \\
&\triangleq e^{-j\omega_k\Delta}X(k)
\end{aligned}$$

The shift theorem is often expressed in shorthand as

$$\boxed{x(n-\Delta) \longleftrightarrow e^{-j\omega_k\Delta}X(\omega_k).}$$

The shift theorem says that a *delay* in the time domain corresponds to a *linear phase term* in the frequency domain. More specifically, a delay of Δ samples in the time waveform corresponds to the linear phase term $e^{-j\omega_k\Delta}$ multiplying the spectrum, where $\omega_k \triangleq 2\pi k/N$.[13] Note that spectral magnitude is unaffected by a linear phase term. That is, $\left|e^{-j\omega_k\Delta}X(k)\right| = |X(k)|$.

Linear Phase Terms

The reason $e^{-j\omega_k\Delta}$ is called a *linear phase term* is that its phase is a linear function of frequency:

$$\angle e^{-j\omega_k\Delta} = -\Delta \cdot \omega_k$$

[13]To consider ω_k as radians per second instead of radians per sample, just replace Δ by ΔT so that the delay is in seconds instead of samples, *i.e.*, $\exp(-j(2\pi k/N)(\Delta)) = \exp(-j(2\pi k/NT)(\Delta T)) = \exp(-j\omega_k\Delta T)$.

Thus, the *slope* of the phase, viewed as a linear function of radian-frequency ω_k, is $-\Delta$. In general, the *time delay in samples* equals *minus the slope of the linear phase term.* If we express the original spectrum in polar form as

$$X(k) = G(k)e^{j\Theta(k)},$$

where G and Θ are the magnitude and phase of X, respectively (both real), we can see that a linear phase term only modifies the spectral phase $\Theta(k)$:

$$e^{-j\omega_k\Delta}X(k) \triangleq e^{-j\omega_k\Delta}G(k)e^{j\Theta(k)} = G(k)e^{j[\Theta(k)-\omega_k\Delta]}$$

where $\omega_k \triangleq 2\pi k/N$. A positive time delay (waveform shift to the right) adds a *negatively sloped* linear phase to the original spectral phase. A negative time delay (waveform shift to the left) adds a *positively sloped* linear phase to the original spectral phase. If we seem to be belaboring this relationship, it is because it is one of the most useful in practice.

Linear Phase Signals

In practice, a *signal* may be said to be *linear phase* when its phase is of the form

$$\Theta(\omega_k) = -\Delta \cdot \omega_k \pm \pi I(\omega_k),$$

where Δ is any real constant (usually an integer), and $I(\omega_k)$ is an *indicator function* which takes on the values 0 or 1 over the points ω_k, $k = 0, 1, 2, \ldots, N-1$. An important class of examples is when the signal is regarded as a filter *impulse response.*[14] What all such signals have in common is that they are *symmetric* about the time $n = \Delta$ in the time domain (as we will show on the next page). Thus, the term "linear phase signal" often really means "a signal whose phase is linear between $\pm\pi$ discontinuities."

Zero Phase Signals

A *zero-phase signal* is thus a linear-phase signal for which the phase-slope Δ is zero. As mentioned above (in §7.4.3), it would be more precise to say "0-or-π-phase signal" instead of "zero-phase signal". Another better term is "zero-centered signal", since every real (even) spectrum corresponds to an even (real) signal. Of course, a zero-centered symmetric signal is simply an *even* signal, by definition. Thus, a "zero-phase signal" is more precisely termed an "even signal".

---

[14]See §8.3 on page 185 and/or Book II [65] of the music signal processing book series for an introduction to digital filters.

Application of the Shift Theorem to FFT Windows

In practical spectrum analysis, we most often use the *Fast Fourier Transform*[15] (FFT) together with a *window function* $w(n), n = 0, 1, 2, \ldots, N-1$. As discussed further in Chapter 8, windows are normally positive ($w(n) > 0$), symmetric about their midpoint, and look pretty much like a "bell curve." A window multiplies the signal x being analyzed to form a *windowed signal* $x_w(n) = w(n)x(n)$, or $x_w = w \cdot x$, which is then analyzed using an FFT. The window serves to *taper* the data segment gracefully to zero, thus eliminating spectral distortions due to suddenly cutting off the signal in time. Windowing is thus appropriate when x is a short section of a longer signal (not a period or whole number of periods from a periodic signal).

Theorem: Real symmetric FFT windows are *linear phase.*

Proof: Let $w(n)$ denote the window samples for $n = 0, 1, 2, \ldots, M-1$. Since the window is symmetric, we have $w(n) = w(M-1-n)$ for all n. When M is odd, there is a sample at the midpoint at time $n = (M-1)/2$. The midpoint can be translated to the time origin to create an even signal. As established on page 148, the DFT of a real and even signal is real and even. By the shift theorem, the DFT of the original symmetric window is a real, even spectrum multiplied by a *linear phase term*, yielding a spectrum having a phase that is linear in frequency with possible discontinuities of $\pm\pi$ radians. Thus, all odd-length real symmetric signals are "linear phase", including FFT windows.

When M is even, the window midpoint at time $n = (M-1)/2$ lands half-way between samples, so we cannot simply translate the window to zero-centered form. However, we can still factor the window spectrum $W(\omega_k)$ into the product of a linear phase term $\exp[-\omega_k(M-1)/2]$ and a real spectrum (verify this as an exercise), which satisfies the definition of a linear phase signal.

[15]An FFT is just a fast implementation of the DFT. See Appendix A for details and pointers.

7.4.5 Convolution Theorem

Theorem: For any $x, y \in \mathbf{C}^N$,

$$\boxed{x \circledast y \longleftrightarrow X \cdot Y.}$$

Proof:

$$\begin{aligned}
\text{DFT}_k(x \circledast y) &\triangleq \sum_{n=0}^{N-1} (x \circledast y)_n e^{-j2\pi nk/N} \\
&\triangleq \sum_{n=0}^{N-1} \sum_{m=0}^{N-1} x(m)y(n-m)e^{-j2\pi nk/N} \\
&= \sum_{m=0}^{N-1} x(m) \sum_{n=0}^{N-1} \underbrace{y(n-m)e^{-j2\pi nk/N}}_{e^{-j2\pi mk/N}Y(k)} \\
&= \left(\sum_{m=0}^{N-1} x(m)e^{-j2\pi mk/N}\right) Y(k) \quad \text{(by the Shift Theorem)} \\
&\triangleq X(k)Y(k)
\end{aligned}$$

This is perhaps the most important single Fourier theorem of all. It is the basis of a large number of FFT applications. Since an FFT provides a fast Fourier transform, it also provides *fast convolution*, thanks to the convolution theorem. It turns out that using an FFT to perform convolution is really more efficient in practice only for reasonably long convolutions, such as $N > 100$. For much longer convolutions, the savings become enormous compared with "direct" convolution. This happens because direct convolution requires on the order of N^2 operations (multiplications and additions), while FFT-based convolution requires on the order of $N \lg(N)$ operations, where $\lg(N)$ denotes the logarithm-base-2 of N (see §A.1.2 on page 199 for an explanation).

The simple matlab example in Fig. 7.13 illustrates how much faster convolution can be performed using an FFT.[16] We see that for a length $N = 1024$ convolution, the `fft` function is approximately 300 times faster in Octave, and 30 times faster in Matlab. (The `conv` routine is much faster in Matlab, even though it is a built-in function in both cases.)

[16] These results were obtained using the program `Octave`, version 2.9.9, running on a Linux PC with a 2.8GHz Pentium CPU, and `Matlab`, version 5.2, running on a Windows PC with an 800MHz Athlon CPU.

```
N = 1024;           % FFT much faster at this length
t = 0:N-1;          % [0,1,2,...,N-1]
h = exp(-t);        % filter impulse reponse
H = fft(h);         % filter frequency response
x = ones(1,N);      % input = dc (any signal will do)
Nrep = 100;         % number of trials to average
t0 = clock;         % latch the current time
for i=1:Nrep, y = conv(x,h); end      % Direct convolution
t1 = etime(clock,t0)*1000; % elapsed time in msec
t0 = clock;
for i=1:Nrep, y = ifft(fft(x) .* H); end % FFT convolution
t2 = etime(clock,t0)*1000;
disp(sprintf([...
    'Average direct-convolution time = %0.2f msec\n',...
    'Average FFT-convolution time = %0.2f msec\n',...
    'Ratio = %0.2f (Direct/FFT)'],...
    t1/Nrep,t2/Nrep,t1/t2));

% ================== EXAMPLE RESULTS ==================

Octave:
Average direct-convolution time = 69.49 msec
Average FFT-convolution time = 0.23 msec
Ratio = 296.40 (Direct/FFT)

Matlab:
Average direct-convolution time = 15.73 msec
Average FFT-convolution time = 0.50 msec
Ratio = 31.46 (Direct/FFT)
```

Figure 7.13: Matlab/Octave program for comparing the speed of direct convolution with that of FFT convolution.

A similar program produced the results for different FFT lengths shown in Table 7.1.[17] In this software environment, the `fft` function is faster starting with length $2^6 = 64$, and it is never significantly slower at short lengths, where "calling overhead" dominates.

M	Direct	FFT	Ratio
1	0.07	0.08	0.91
2	0.08	0.08	0.92
3	0.08	0.08	0.94
4	0.09	0.10	0.97
5	0.12	0.12	0.96
6	0.18	0.12	1.44
7	0.39	0.15	2.67
8	1.10	0.21	5.10
9	3.83	0.31	12.26
10	15.80	0.47	33.72
11	50.39	1.09	46.07
12	177.75	2.53	70.22
13	709.75	5.62	126.18
14	4510.25	17.50	257.73
15	19050.00	72.50	262.76
16	316375.00	440.50	718.22

Table 7.1: Direct versus FFT convolution times in milliseconds (convolution length $= 2^M$) using Matlab 5.2 on an 800 MHz Athlon Windows PC.

A table similar to Table 7.1 in Strum and Kirk [76, p. 521], based on the number of real multiplies, finds that the `fft` is faster starting at length $2^7 = 128$, and that direct convolution is significantly faster for very short convolutions (*e.g.*, 16 operations for a direct length-4 convolution, versus 176 for the `fft` function).

See Appendix A for further discussion of FFT algorithms and their applications.

[17]These results were obtained using `Matlab` v5.2 running on a Windows PC with an 800MHz Athlon CPU.

7.4.6 Dual of the Convolution Theorem

The *dual*[18] of the convolution theorem says that *multiplication in the time domain is convolution in the frequency domain*:

Theorem:

$$\boxed{x \cdot y \longleftrightarrow \frac{1}{N} X \circledast Y}$$

Proof: The steps are the same as in the convolution theorem.

This theorem also bears on the use of *FFT windows*. It implies that *windowing in the time domain* corresponds to *smoothing in the frequency domain*. That is, the spectrum of $w \cdot x$ is simply X filtered by W, or, $W \circledast X$. This smoothing reduces *sidelobes* associated with the *rectangular window*, which is the window one is using implicitly when a data frame x is considered *time limited* and therefore eligible for "windowing" (and zero-padding). See Chapter 8 and Book IV [67] for further discussion.

7.4.7 Correlation Theorem

Theorem: For all $x, y \in \mathbf{C}^N$,

$$\boxed{x \star y \longleftrightarrow \overline{X} \cdot Y}$$

where the correlation operation '$\star$' was defined in §7.2.5 on page 129.

Proof:

$$\begin{aligned}
(x \star y)_n &\triangleq \sum_{m=0}^{N-1} \overline{x(m)} y(n+m) \\
&= \sum_{m=0}^{N-1} \overline{x(-m)} y(n-m) \qquad (m \leftarrow -m) \\
&= (\text{FLIP}(\overline{x}) \circledast y)_n \\
&\longleftrightarrow \overline{X} \cdot Y
\end{aligned}$$

The last step follows from the convolution theorem and the result $\text{FLIP}(\overline{x}) \longleftrightarrow \overline{X}$ from §7.4.2 on page 145. Also, the summation range in the second line is equivalent to the range $[N-1, 0]$ because all indexing is modulo N.

[18]The *dual* of a Fourier operation is obtained by interchanging time and frequency.

7.4.8 Power Theorem

Theorem: For all $x, y \in \mathbf{C}^N$,

$$\boxed{\langle x, y\rangle = \frac{1}{N}\langle X, Y\rangle .}$$

Proof:

$$\begin{aligned}\langle x, y\rangle &\isdef \sum_{n=0}^{N-1} x(n)\overline{y(n)} = (y \star x)_0 = \operatorname{DFT}_0^{-1}(\overline{Y}\cdot X)\\ &= \frac{1}{N}\sum_{k=0}^{N-1} X(k)\overline{Y(k)} \isdef \frac{1}{N}\langle X, Y\rangle .\end{aligned}$$

As mentioned in §5.8 on page 74, physical *power* is *energy per unit time*.[19] For example, when a force produces a motion, the power delivered is given by the *force* times the *velocity* of the motion. Therefore, if $x(n)$ and $y(n)$ are in physical units of force and velocity (or any analogous quantities such as voltage and current, etc.), then their product $x(n)y(n) \isdef f(n)v(n)$ is proportional to the *power per sample* at time n, and $\langle f, v\rangle$ becomes proportional to the total *energy* supplied (or absorbed) by the driving force. By the power theorem, $F(k)\overline{V(k)}/N$ can be interpreted as the *energy per bin* in the DFT, or *spectral power*, *i.e.*, the energy associated with a spectral *band* of width $2\pi/N$.[20]

Normalized DFT Power Theorem

Note that the power theorem would be more elegant if the DFT were defined as the coefficient of projection onto the *normalized DFT sinusoids*

$$\tilde{s}_k(n) \isdef \frac{s_k(n)}{\sqrt{N}}.$$

[19]Signal processing with a physical interpretation is addressed in Book III [66] of the music signal processing series.

[20]You might be concerned about what it means when $F(k)\overline{V(k)}/N$ has an *imaginary* part. This happens when the force $f(t)$ drives a so-called "reactive" load such as a mass or spring. For the details, see Book III [66] or any text on classical network theory or circuit theory. When the driving force $f(t)$ works against a real resistance $R > 0$, then $f(t) = Rv(t)$ (see §F.3 on page 250), so that the power is $f(t)v(t) = Rv^2(t) = f^2(t)/R$, and in the frequency domain, $F(\omega) = RV(\omega)$ so that the spectral power is proportional to $F(\omega)\overline{V(\omega)} = R\left|V(\omega)\right|^2 = \left|F(\omega)\right|^2/R$. Thus, in the case of a real driving-point impedance R, the spectral power is always real and nonnegative.

That is, for the *normalized DFT* (§6.10 on page 109), the power theorem becomes simply

$$\langle x, y\rangle = \langle \tilde{X}, \tilde{Y}\rangle \qquad \text{(Normalized DFT case).}$$

We see that the power theorem expresses the invariance of the inner product between two signals in the time and frequency domains. If we think of the inner product *geometrically*, as in Chapter 5, then this result is expected, because x and $\tilde{X}$ are merely coordinates of the same geometric object (a signal) relative to two different sets of basis signals (the shifted impulses and the normalized DFT sinusoids).

7.4.9 Rayleigh Energy Theorem (Parseval's Theorem)

Theorem: For any $x \in \mathbf{C}^N$,

$$\boxed{\|\, x \,\|^2 = \frac{1}{N} \|\, X \,\|^2 .}$$

I.e.,

$$\boxed{\sum_{n=0}^{N-1} |x(n)|^2 = \frac{1}{N} \sum_{k=0}^{N-1} |X(k)|^2 .}$$

Proof: This is a special case of the power theorem.

Note that again the relationship would be cleaner ($\|\, x \,\| = \|\, \tilde{X} \,\|$) if we were using the *normalized DFT*.

7.4.10 Stretch Theorem (Repeat Theorem)

Theorem: For all $x \in \mathbf{C}^N$,

$$\boxed{\text{Stretch}_L(x) \longleftrightarrow \text{Repeat}_L(X).}$$

Proof: Recall the stretch operator:

$$\text{Stretch}_{L,m}(x) \triangleq \begin{cases} x(m/L), & m/L = \text{integer} \\ 0, & m/L \neq \text{integer} \end{cases}$$

Let $y \triangleq \text{Stretch}_L(x)$, where $y \in \mathbf{C}^M$, $M = LN$. Also define the new denser frequency grid associated with length M by $\omega'_k \triangleq 2\pi k/M$, and define

$\omega_k = 2\pi k/N$ as usual. Then

$$Y(k) \triangleq \sum_{m=0}^{M-1} y(m)e^{-j\omega'_k m} = \sum_{n=0}^{N-1} x(n)e^{-j\omega'_k nL} \qquad (n \triangleq m/L).$$

But

$$\omega'_k L \triangleq \frac{2\pi k}{M} L = \frac{2\pi k}{N} = \omega_k.$$

Thus, $Y(k) = X(k)$, and by the modulo indexing of X, L copies of X are generated as k goes from 0 to $M - 1 = LN - 1$.

7.4.11 Downsampling Theorem (Aliasing Theorem)

Theorem: For all $x \in \mathbf{C}^N$,

$$\boxed{\text{DOWNSAMPLE}_L(x) \longleftrightarrow \frac{1}{L}\text{ALIAS}_L(X).}$$

Proof: Let $k' \in [0, M-1]$ denote the frequency index in the aliased spectrum, and let $Y(k') \triangleq \text{ALIAS}_{L,k'}(X)$. Then Y is length $M = N/L$, where L is the downsampling factor. We have

$$\begin{aligned}
Y(k') &\triangleq \text{ALIAS}_{L,k'}(X) \triangleq \sum_{l=0}^{L-1} X(k'+lM), \quad k' = 0,1,2,\ldots,M-1 \\
&\triangleq \sum_{l=0}^{L-1}\sum_{n=0}^{N-1} x(n)e^{-j2\pi(k'+lM)n/N} \\
&\triangleq \sum_{n=0}^{N-1} x(n)e^{-j2\pi k'n/N} \sum_{l=0}^{L-1} e^{-j2\pi lnM/N}.
\end{aligned}$$

Since $M/N = 1/L$, the sum over l becomes

$$\sum_{l=0}^{L-1} \left[e^{-j2\pi n/L}\right]^l = \frac{1-e^{-j2\pi n}}{1-e^{-j2\pi n/L}} = \begin{cases} L, & n = 0 \,(\text{mod } L) \\ 0, & n \neq 0 \,(\text{mod } L) \end{cases}$$

using the closed form expression for a geometric series derived in §6.1 on page 99. We see that the sum over L effectively *samples* x every L samples. This can be expressed in the previous formula by defining $m \triangleq n/L$ which ranges only over the nonzero samples:

$$\begin{aligned}
\text{ALIAS}_{L,k'}(X) &= \sum_{n=0}^{N-1} x(n)e^{-j2\pi k'n/N} \sum_{l=0}^{L-1} e^{-j2\pi ln/L} \\
&= L \sum_{m=0}^{N/L-1} x(mL)e^{-j2\pi k'(mL)/N} \qquad (m \stackrel{\Delta}{=} n/L) \\
&\stackrel{\Delta}{=} L \sum_{m=0}^{M-1} x(mL)e^{-j2\pi k'm/M} \\
&\stackrel{\Delta}{=} L \sum_{m=0}^{M-1} \text{DOWNSAMPLE}_{L,m}(x)e^{-j2\pi k'm/M} \\
&\stackrel{\Delta}{=} L \cdot \text{DFT}_{k'}(\text{DOWNSAMPLE}_L(x))
\end{aligned}$$

Since the above derivation also works in reverse, the theorem is proved.

An illustration of aliasing in the frequency domain is shown in Fig. 7.12 on page 141.

Illustration of the Downsampling/Aliasing Theorem in Matlab

```
>> N=4;
>> x = 1:N;
>> X = fft(x);
>> x2 = x(1:2:N);
>> fft(x2)                          % FFT(Downsample(x,2))
ans =
    4   -2
>> (X(1:N/2) + X(N/2 + 1:N))/2      % (1/2) Alias(X,2)
ans =
    4   -2
```

7.4.12 Zero Padding Theorem (Spectral Interpolation)

A fundamental tool in practical spectrum analysis is *zero padding*. This theorem shows that zero padding in the time domain corresponds to *ideal interpolation in the frequency domain* (for time-limited signals):

Theorem: For any $x \in \mathbf{C}^N$

$$\boxed{\text{ZEROPAD}_{LN}(x) \longleftrightarrow \text{INTERP}_L(X)}$$

where ZEROPAD() was defined in Eq. (7.4) on page 131, followed by the definition of INTERP().

Proof: Let $M = LN$ with $L \geq 1$. Then

$$\begin{aligned}\text{DFT}_{M,k'}(\text{ZeroPad}_M(x)) &= \sum_{m=0}^{M-1} x(m)e^{-j2\pi mk'/M} \triangleq \langle x, s_{\omega_{k'}} \rangle \\ &\triangleq X(\omega_{k'}) = \text{Interp}_{L,k'}(X).\end{aligned}$$

Thus, this theorem follows directly from the definition of the ideal interpolation operator INTERP(). See §8.1.3 on page 170 for an example of zero-padding in spectrum analysis.

7.4.13 Periodic Interpolation (Spectral Zero Padding)

The *dual* of the zero-padding theorem states formally that *zero padding in the frequency domain* corresponds to *periodic interpolation* in the time domain:

Definition: For all $x \in \mathbf{C}^N$ and any integer $L \geq 1$,

$$\boxed{\text{PerInterp}_L(x) \triangleq \text{IDFT}(\text{ZeroPad}_{LN}(X))} \tag{7.7}$$

where zero padding is defined in §7.2.7 and illustrated in Figure 7.7 on page 132. In other words, zero-padding a DFT by the factor L in the frequency domain (by inserting $N(L-1)$ zeros at bin number $k = N/2$ corresponding to the *folding frequency*[21]) gives rise to "periodic interpolation" by the factor L in the time domain. It is straightforward to show that the interpolation kernel used in periodic interpolation is an *aliased sinc function*, that is, a sinc function $\sin(\pi n/L)/(\pi n/L)$ that has been time-aliased on a block of length NL. Such an aliased sinc function is of course periodic with period NL samples. See Appendix D for a discussion of *ideal* bandlimited interpolation, in which the interpolating sinc function is not aliased.

Periodic interpolation is ideal for signals that are *periodic* in N samples, where N is the DFT length. For non-periodic signals, which is almost always the case in practice, bandlimited interpolation should be used instead (Appendix D).

Relation to Stretch Theorem

It is instructive to interpret the periodic interpolation theorem in terms of the stretch theorem, $\text{Stretch}_L(x) \longleftrightarrow \text{Repeat}_L(X)$. To do this, it is convenient to define a "zero-centered rectangular window" operator:

[21]The *folding frequency* is defined as half the sampling rate $f_s/2$. It may also be called the *Nyquist limit*. The *Nyquist rate*, on the other hand, means the sampling rate, not half the sampling rate.

Definition: For any $X \in \mathbf{C}^N$ and any odd integer $M < N$ we define the *length M even rectangular windowing operation* by

$$\text{CHOP}_{M,k}(X) \triangleq \begin{cases} X(k), & -\frac{M-1}{2} \le k \le \frac{M-1}{2} \\ 0, & \frac{M+1}{2} \le |k| \le \frac{N}{2}. \end{cases}$$

Thus, this "zero-phase rectangular window," when applied to a spectrum X, sets the spectrum to zero everywhere outside a zero-centered interval of M samples. Note that $\text{CHOP}_M(X)$ is the *ideal lowpass filtering operation in the frequency domain.* The "cut-off frequency" is $\omega_c = 2\pi[(M-1)/2]/N$ radians per sample. For even M, we allow $X(-M/2)$ to be "passed" by the window, but in our usage (below), this sample should always be zero anyway. With this notation defined we can efficiently restate *periodic interpolation* in terms of the $\text{STRETCH}()$ operator:

Theorem: When $x \in \mathbf{C}^N$ consists of one or more periods from a *periodic* signal $x' \in \mathbf{C}^\infty$,

$$\boxed{\text{PERINTERP}_L(x) = \text{IDFT}(\text{CHOP}_N(\text{DFT}(\text{STRETCH}_L(x)))).}$$

In other words, ideal periodic interpolation of one period of x by the integer factor L may be carried out by first stretching x by the factor L (inserting $L-1$ zeros between adjacent samples of x), taking the DFT, applying the ideal lowpass filter as an N-point rectangular window in the frequency domain, and performing the inverse DFT.

Proof: First, recall that $\text{STRETCH}_L(x) \leftrightarrow \text{REPEAT}_L(X)$. That is, stretching a signal by the factor L gives a new signal $y = \text{STRETCH}_L(x)$ which has a spectrum Y consisting of L copies of X repeated around the unit circle. The "baseband copy" of X in Y can be defined as the N-sample sequence centered about frequency zero. Therefore, we can use an "ideal filter" to "pass" the baseband spectral copy and zero out all others, thereby converting $\text{REPEAT}_L(X)$ to $\text{ZEROPAD}_{LN}(X)$. *I.e.*,

$$\text{CHOP}_N(\text{REPEAT}_L(X)) = \text{ZEROPAD}_{LN}(X) \longleftrightarrow \text{INTERP}_L(x).$$

The last step is provided by the zero-padding theorem (§7.4.12).

Bandlimited Interpolation of Time-Limited Signals

The previous result can be extended toward bandlimited interpolation of $x \in \mathbf{C}^{N_x}$ which includes all nonzero samples from an *arbitrary* time-limited signal $x' \in \mathbf{C}^\infty$ (*i.e.*, going beyond the interpolation of only periodic bandlimited signals given one or more periods $x \in \mathbf{C}^N$) by

1. replacing the rectangular window $\text{CHOP}_N()$ with a *smoother spectral window* $H(\omega)$, and

2. using extra zero-padding in the time domain to convert the *cyclic* convolution between $\text{STRETCH}_L(x)$ and h into an *acyclic* convolution between them (recall §7.2.4 on page 121).

The smoother spectral window H can be thought of as the *frequency response* of the FIR[22] filter h used as the bandlimited interpolation kernel in the time domain. The number of zeros needed in the zero-padding of x in the time domain is simply length of h minus 1, and the number of zeros to be appended to h is the length of $\text{STRETCH}_L(x)$ minus 1. With this much zero-padding, the cyclic convolution of x and h implemented using the DFT becomes equivalent to acyclic convolution, as desired for the time-limited signals x and h. Thus, if N_x denotes the nonzero length of x, then the nonzero length of $\text{STRETCH}_L(x)$ is $L(N_x - 1) + 1$, and we require the DFT length to be $N \geq L(N_x - 1) + N_h$, where N_h is the filter length. In operator notation, we can express bandlimited sampling-rate up-conversion by the factor L for time-limited signals $x \in \mathbf{C}^{N_x}$ by

$$\boxed{\text{INTERP}_L(x) \approx \text{IDFT}(H \cdot (\text{DFT}(\text{ZEROPAD}_N(\text{STRETCH}_L(x)))).} \tag{7.8}$$

The approximation symbol '$\approx$' approaches equality as the spectral window H approaches $\text{CHOP}_{N_x}([1, \ldots, 1])$ (the frequency response of the ideal lowpass filter passing only the original spectrum X), while at the same time allowing no time aliasing (convolution remains acyclic in the time domain).

Equation (7.8) can provide the basis for a high-quality sampling-rate conversion algorithm. Arbitrarily long signals can be accommodated by breaking them into segments of length N_x, applying the above algorithm to each block, and summing the up-sampled blocks using *overlap-add*. That is, the lowpass filter h "rings" into the next block and possibly beyond (or even into both adjacent time blocks when h is not causal), and this ringing must be summed into all affected adjacent blocks. Finally, the filter H can "window away" more than the top $L-1$ copies of X in Y, thereby preparing the time-domain signal for *downsampling*, say by $M \in \mathbf{Z}$:

$$\boxed{\text{INTERP}_{L/M}(x) \approx \text{DOWNSAMPLE}_M(\text{IDFT}(H \cdot (\text{DFT}(\text{ZEROPAD}_N(\text{STRETCH}_L(x))))))}$$

where now the lowpass filter frequency response H must be close to zero for all $|\omega_k| \geq \pi / \max(L, M)$. While such a sampling-rate conversion algorithm can be made more efficient by using an FFT in place of the DFT (see

[22]FIR stands for "Finite Impulse Response." Digital filtering concepts and terminology are introduced in §8.3 on page 185 and more completely in Book II [65] of the music signal processing series.

Appendix A), it is not necessarily the most efficient algorithm possible. This is because (1) $M-1$ out of M output samples from the IDFT need not be computed at all, and (2) $\text{STRETCH}_L(x)$ has many zeros in it which do not need explicit handling. For an introduction to time-domain sampling-rate conversion (bandlimited interpolation) algorithms which take advantage of points (1) and (2) in this paragraph, see, *e.g.*, Appendix D and [69].

7.5 DFT Theorem Problems

1. The notation

$$x = y\,(\text{mod } N)$$

(which is read "x equals y modulo N") means there exists some integer k such that $x = y + kN$. Similarly, the notation "x mod N" means $x + kN$ with k chosen to give a result between 0 and $N-1$. In the problems below, find $k \in \mathcal{Z}$ for each case:

   (a) $N = 0\,(\text{mod } N)$

   (b) $-1 = (N-1)\,(\text{mod } N)$

   (c) $X/2 = Y\,(\text{mod } N) \Leftrightarrow X = 2Y\,(\text{mod } 2N)$
   [Hint: See what happens to the "kN" term.]

2. Show that when N is even, the DFT summation can be split into two sums over the odd and even indices of the input signal to obtain

$$\begin{aligned} X(\omega_k) &= \text{DFT}_{\frac{N}{2},k}\{\text{DOWNSAMPLE}_2(x)\} \\ &\quad + W_N^k \cdot \text{DFT}_{\frac{N}{2},k}\{\text{DOWNSAMPLE}_2[\text{SHIFT}_1(x)]\} \end{aligned}$$

In other words, a length N DFT can be computed using two length $N/2$ DFTs. The splitting into sums over even and odd time indices is called *decimation in time.*

3. Show that $\text{DFT}(\text{DFT}(y)) = N \cdot \text{FLIP}(y)$, where $\text{FLIP}_n(y) \triangleq y[-n] = y[N-n]$.

4. A sequence $x[n]$ is said to be *even* if $x[-n] = x[n]$, and *odd* if $x[-n] = -x[n], n = 0, 1, 2, \ldots, N-1$, where indexing is carried out modulo N (i.e., $-n$ (mod N) $= N - n$ for $0 \leq n < N$). Show that every signal $x[n]$, $n = 0, 1, \ldots, N-1$, can be uniquely expressed as $x[n] = x_e[n] + x_o[n]$, where x_e is even and x_o is odd.

5. Find the even and odd parts of the following sequences:

$$\begin{array}{ll} \text{a)} & [0,1,2,1] \\ \text{b)} & [0,1,0,-1] \\ \text{c)} & [0,1,2,3] \\ \text{d)} & [0,0,0,0] \\ \text{e)} & x[n]=n^2, n=0,1,\ldots,N-1 \end{array}$$

6. A spectrum $X[k]$ is said to be *Hermitian* if $X[-k] = X^*[k]$, i.e., its real part is *even* and its imaginary part is *odd*. Given a Hermitian spectrum, what are the symmetry properties of its magnitude and phase?

7. If $Y[k]$ denotes the kth element of the length N DFT of y, show that:

$$\begin{array}{lll} \text{a)} & \text{im}\{y\}=0 \Longleftrightarrow Y[k]=\overline{Y[N-k]} & [\text{DFT}\{\text{real}\} \text{ is } \textit{Hermitian}] \\ \text{b)} & \text{re}\{y\}=0 \Longleftrightarrow Y[k]=-\overline{Y[N-k]} & [\textit{anti-Hermitian}] \\ \text{c)} & y \text{ even} \Longleftrightarrow Y \text{ even} & \\ \text{d)} & y \text{ odd} \Longleftrightarrow Y \text{ odd} & \\ \text{e)} & y \text{ real, even} \Longleftrightarrow Y \text{ real,even} & \\ \text{f)} & y \text{ real, odd} \Longleftrightarrow Y \text{ imag, odd} & \\ \text{g)} & y \text{ imag, even} \Longleftrightarrow Y \text{ imag, even} & \\ \text{h)} & y \text{ imag, odd} \Longleftrightarrow Y \text{ real, odd} & \end{array}$$

8. Let $y = [0, j, 2, 3]$. Find

   (a) $\text{DFT}_k(y)$

   (b) $\text{DFT}_k(\text{DFT}(y))$

   (c) $\text{DFT}_k(\text{DFT}(y + \frac{1}{2}\text{FLIP}(y)))$ [Hint: Linearity saves time]

   (d) $\text{DFT}_k(y + \overline{y})$ [Answer is conjugate-symmetric]

   (e) $\text{DFT}_k(\text{re}\{y\})$

   (f) $\text{DFT}_k([0, j, 2, 3, 0, 0, 0, 0])$ [How is answer related to a)?]

   (g) $\text{DFT}_k([1, j, 2, 3, 0, 0, 0, 0])$ [This should be easy]

   (h) $\text{DFT}_k([0, 1, 0, 0])$

   (i) $\text{DFT}_k([3, 0, j, 2]) = \text{DFT}_k(\text{SHIFT}_1(y))$

   (j) $\text{DFT}_k([j, 3]) = \text{DFT}_k(\text{DOWNSAMPLE}_2(y))$

   (k) $\text{DFT}_k([0, 0, j, 0, 2, 0, 3, 0]) = \text{DFT}_k(\text{STRETCH}_2(y))$

   (l) $\text{DFT}_k(x)$, $x[n] = \cos[2\pi n/8]$, $n = 0, 1, \ldots, 7$

Chapter 8

Example Applications of the DFT

This chapter gives a start on some applications of the DFT. First, we work through a progressive series of *spectrum analysis* examples using an efficient implementation of the DFT in Matlab or Octave. The various Fourier theorems provide a "thinking vocabulary" for understanding elements of spectral analysis. Next, the basics of *linear systems theory* are presented, relying heavily on the convolution theorem and properties of complex numbers. Finally, some applications of the DFT in *statistical signal processing* are introduced, including cross-correlation, matched filtering, system identification, power spectrum estimation, and coherence function measurement. A side topic in this chapter is practical usage of matlab for signal processing, including display of signals and spectra.

Why a DFT is usually called an FFT in practice

Practical implementations of the DFT are usually based on one of the Cooley-Tukey "Fast Fourier Transform" (FFT) algorithms [14].[1] For this reason, the matlab DFT function is called '`fft`', and the actual algorithm used depends primarily on the transform length N.[2] The fastest FFT algorithms generally occur when N is a power of 2. In practical audio signal processing, we routinely zero-pad our FFT input buffers to the next power of 2 in length (thereby interpolating our spectra somewhat) in order to

[1]While a length N DFT requires approximately N^2 arithmetic operations, a Cooley-Tukey FFT requires closer to $N \lg N = NK$ operations when $N = 2^K$ is a power of 2, where $\lg N$ denotes the log-base-2 of N. Appendix A provides an introduction to Cooley-Tukey FFT algorithms in §A.1 on page 198.

[2]Say "`doc fft`" in Matlab for an overview of how a specific FFT algorithm is chosen.

enjoy the power-of-2 speed advantage. Finer spectral sampling is a typically welcome side benefit of increasing N to the next power of 2. Appendix A provides a short overview of some of the better known FFT algorithms, and some pointers to literature and online resources.

8.1 Spectrum Analysis of a Sinusoid: Windowing, Zero-Padding, and FFT

The examples below give a progression from the most simplistic analysis up to a proper practical treatment. Careful study of these examples will teach you a lot about how spectrum analysis is carried out on real data, and provide opportunities to see the Fourier theorems in action.

8.1.1 FFT of a Simple Sinusoid

Our first example is an FFT of the simple sinusoid

$$x(n) = \cos(\omega_x nT)$$

where we choose $\omega_x = 2\pi(f_s/4)$ (frequency $f_s/4$ Hz) and $T = 1$ (sampling rate f_s set to 1). Since we're using a Cooley-Tukey FFT, the signal length N should be a power of 2 for fastest results. Here is the Matlab code:

```
% Example 1: FFT of a DFT-sinusoid

% Parameters:
N = 64;                % Must be a power of two
T = 1;                 % Set sampling rate to 1
A = 1;                 % Sinusoidal amplitude
phi = 0;               % Sinusoidal phase
f = 0.25;              % Frequency (cycles/sample)
n = [0:N-1];           % Discrete time axis
x = A*cos(2*pi*n*f*T+phi); % Sampled sinusoid
X = fft(x);            % Spectrum

% Plot time data:
figure(1);
subplot(3,1,1);
plot(n,x,'*k');
ni = [0:.1:N-1];       % Interpolated time axis
hold on;
plot(ni,A*cos(2*pi*ni*f*T+phi),'-k'); grid off;
title('Sinusoid at 1/4 the Sampling Rate');
xlabel('Time (samples)');
```

```
ylabel('Amplitude');
text(-8,1,'a)');
hold off;

% Plot spectral magnitude:
magX = abs(X);
fn = [0:1/N:1-1/N];  % Normalized frequency axis
subplot(3,1,2);
stem(fn,magX,'ok'); grid on;
xlabel('Normalized Frequency (cycles per sample))');
ylabel('Magnitude (Linear)');
text(-.11,40,'b)');

% Same thing on a dB scale:
spec = 20*log10(magX); % Spectral magnitude in dB
subplot(3,1,3);
plot(fn,spec,'--ok'); grid on;
axis([0 1 -350 50]);
xlabel('Normalized Frequency (cycles per sample))');
ylabel('Magnitude (dB)');
text(-.11,50,'c)');
cmd = ['print -deps ', '../eps/example1.eps'];
disp(cmd); eval(cmd);
```

The results are shown in Fig. 8.1. The time-domain signal is shown in the upper plot (Fig. 8.1a), both in pseudo-continuous and sampled form. In the middle plot (Fig. 8.1b), we see two peaks in the magnitude spectrum, each at magnitude 32 on a linear scale, located at normalized frequencies $f = 0.25$ and $f = 0.75 = -0.25$. A spectral peak amplitude of $32 = (1/2)64$ is what we expect, since

$$\text{DFT}_k(\cos(\omega_x n)) \stackrel{\Delta}{=} \sum_{n=0}^{N-1} \frac{e^{j\omega_x n} + e^{-j\omega_x n}}{2} e^{-j\omega_k n},$$

and when $\omega_k = \pm\omega_x$, this reduces to

$$\sum_{n=0}^{N-1} \frac{e^{j0n}}{2} = \frac{N}{2}.$$

For $N = 64$ and $\omega_x = 2\pi f_s/4$, this happens at bin numbers $k = 0.25N = 16$ and $k = 0.75N = 48$. However, recall that array indexes in matlab start at 1, so that these peaks will really show up at indexes 17 and 49 in the `magX` array.

The spectrum should be exactly zero at the other bin numbers. How accurately this happens can be seen by looking on a dB scale, as shown in

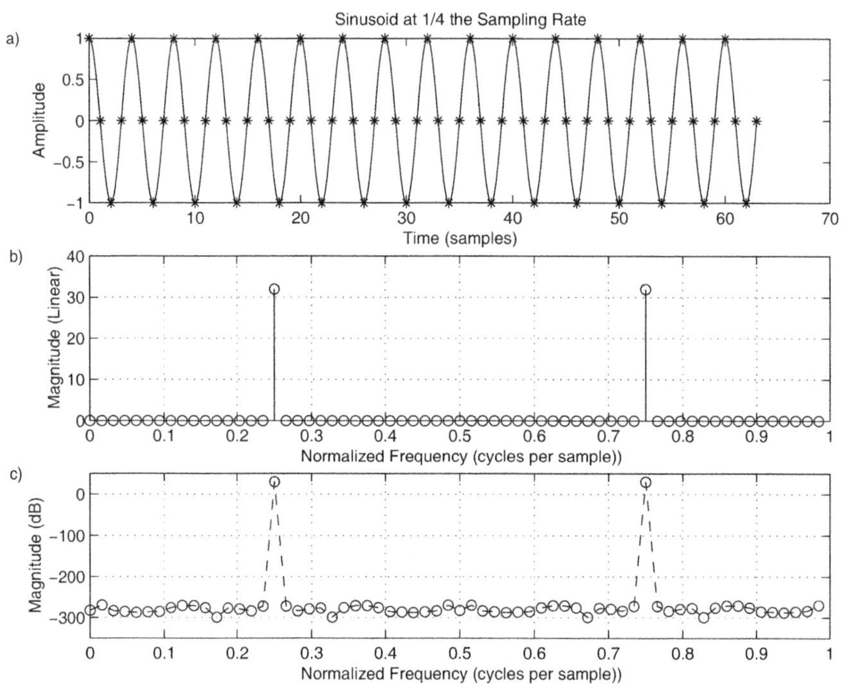

Figure 8.1: Sampled sinusoid at frequency $f = f_s/4$. a) Time waveform. b) Magnitude spectrum. c) DB magnitude spectrum.

Fig. 8.1c. We see that the spectral magnitude in the other bins is on the order of 300 dB lower, which is close enough to zero for audio work (◡̈).

8.1.2 FFT of a Not-So-Simple Sinusoid

Now let's increase the frequency in the above example by one-half of a bin:

```
% Example 2 = Example 1 with frequency between bins

f = 0.25 + 0.5/N;    % Move frequency up 1/2 bin

x = cos(2*pi*n*f*T); % Signal to analyze
X = fft(x);          % Spectrum
...                  % See Example 1 for plots and such
```

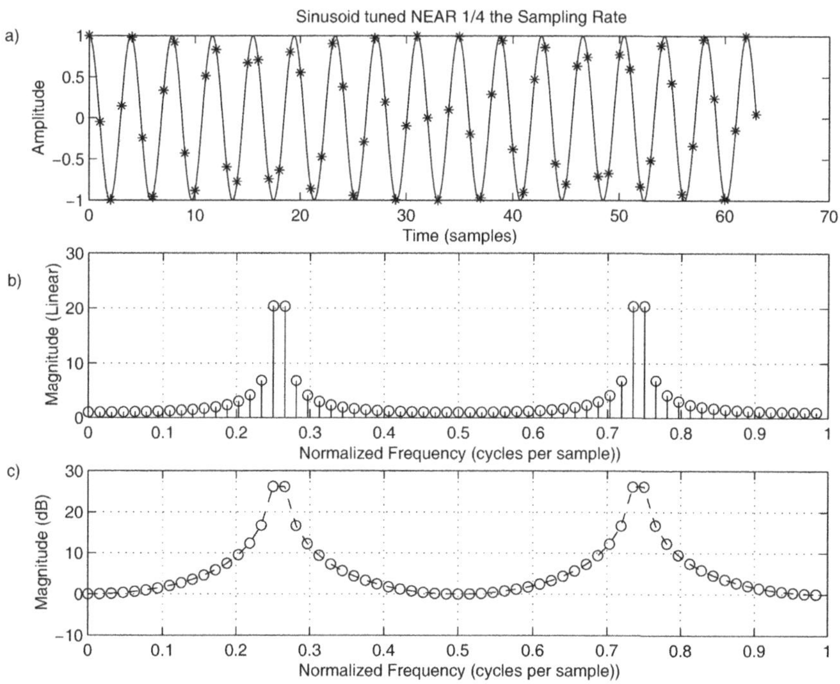

Figure 8.2: Sinusoid at Frequency $f = 0.25 + 0.5/N$. a) Time waveform. b) Magnitude spectrum. c) DB magnitude spectrum.

The resulting magnitude spectrum is shown in Fig. 8.2b and c. At this frequency, we get extensive "spectral leakage" into all the bins. To get an idea of where this is coming from, let's look at the *periodic extension* (§7.1.2) of the time waveform:

```
% Plot the periodic extension of the time-domain signal
plot([x,x],'--ok');
title('Time Waveform Repeated Once');
xlabel('Time (samples)'); ylabel('Amplitude');
```

The result is shown in Fig. 8.3. Note the "glitch" in the middle where the signal begins its forced repetition.

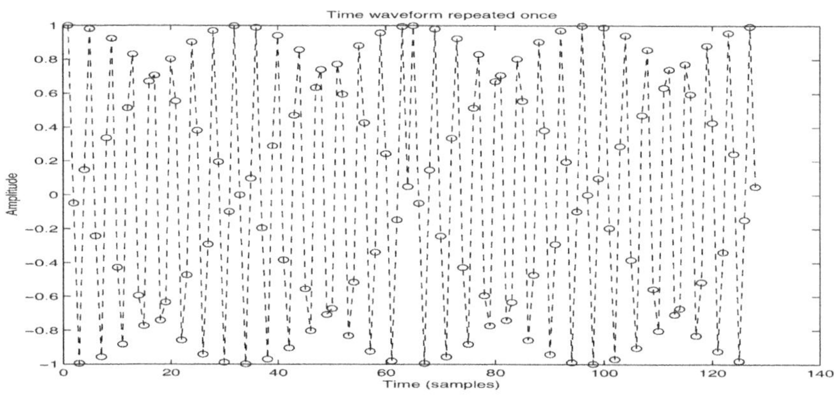

Figure 8.3: Time waveform repeated to show discontinuity introduced by periodic extension (see midpoint).

8.1.3 FFT of a Zero-Padded Sinusoid

Looking back at Fig. 8.2c, we see there are no negative dB values. Could this be right? Could the spectral magnitude at all frequencies be 1 or greater? The answer is no. To better see the true spectrum, let's use *zero padding* in the time domain (§7.2.7) to give *ideal interpolation* (§7.4.12) in the frequency domain:

```
zpf = 8;                % zero-padding factor
x = [cos(2*pi*n*f*T),zeros(1,(zpf-1)*N)]; % zero-padded
X = fft(x);             % interpolated spectrum
magX = abs(X);          % magnitude spectrum
...                     % waveform plot as before
nfft = zpf*N;           % FFT size = new frequency grid size
fni = [0:1.0/nfft:1-1.0/nfft]; % normalized freq axis
subplot(3,1,2);
```

```
% with interpolation, we can use solid lines '-':
plot(fni,magX,'-k'); grid on;
...
spec = 20*log10(magX); % spectral magnitude in dB
% clip below at -40 dB:
spec = max(spec,-40*ones(1,length(spec)));
...                    % plot as before
```

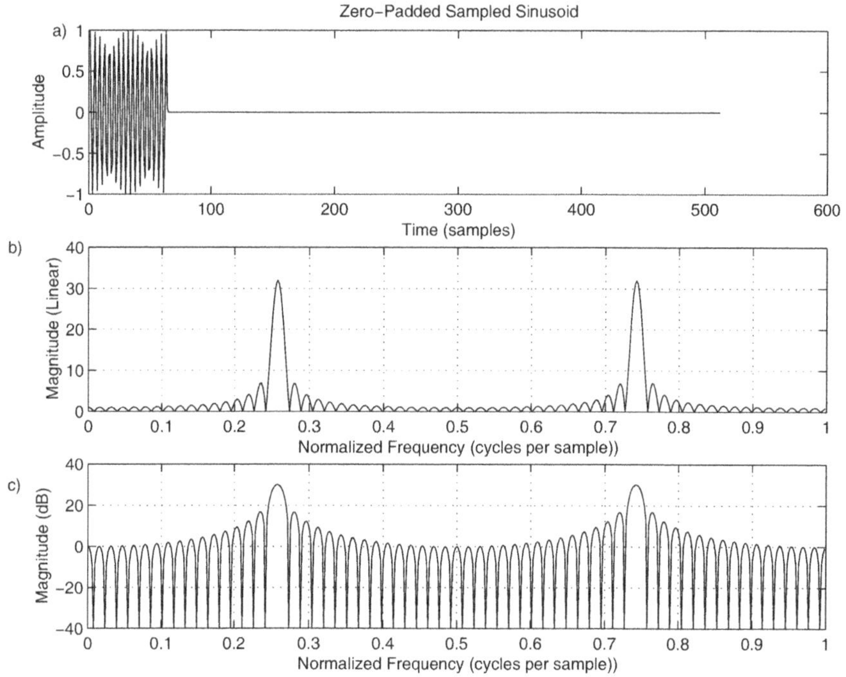

Figure 8.4: Zero-padded sinusoid at frequency $f = 0.25 + 0.5/N$ cycles/sample. a) Time waveform. b) Magnitude spectrum. c) DB magnitude spectrum.

Figure 8.4 shows the zero-padded data (top) and corresponding interpolated spectrum on linear and dB scales (middle and bottom, respectively). We now see that the spectrum has a regular *sidelobe* structure. On the dB scale in Fig. 8.4c, negative values are now visible. In fact, it was desirable to *clip* them at -40 dB to prevent deep nulls from dominating the display by pushing the negative vertical axis limit to -300 dB or more, as in Fig. 8.1c (page 168). This example shows the importance of using zero padding to interpolate spectral displays so that the untrained eye will "fill in" properly between the spectral samples.

8.1.4 Use of a Blackman Window

As Fig. 8.4a suggests, the previous example can be interpreted as using a *rectangular window* to select a finite segment (of length N) from a sampled sinusoid that continues for all time. In practical spectrum analysis, such excerpts are normally analyzed using a window that is *tapered* more gracefully to zero on the left and right. In this section, we will look at using a *Blackman window* [67][3] on our example sinusoid. The Blackman window has good (though suboptimal) characteristics for audio work.

In Octave[4] or the Matlab Signal Processing Toolbox,[5] a Blackman window of length $M = 64$ can be designed very easily:

```
M = 64;
w = blackman(M);
```

Many other standard windows are defined as well, including `hamming`, `hanning`, and `bartlett` windows.

In Matlab without the Signal Processing Toolbox, the Blackman window is readily computed from its mathematical definition:

```
w = .42 - .5*cos(2*pi*(0:M-1)/(M-1)) ...
      + .08*cos(4*pi*(0:M-1)/(M-1));
```

Figure 8.5 shows the Blackman window and its magnitude spectrum on a dB scale. Fig. 8.5c uses the more "physical" frequency axis in which the upper half of the FFT bin numbers are interpreted as negative frequencies. Here is the complete Matlab script for Fig. 8.5:

```
M = 64;
w = blackman(M);
figure(1);
subplot(3,1,1); plot(w,'*'); title('Blackman Window');
xlabel('Time (samples)'); ylabel('Amplitude'); text(-8,1,'a)');

% Also show the window transform:
zpf = 8;                      % zero-padding factor
xw = [w',zeros(1,(zpf-1)*M)]; % zero-padded window
Xw = fft(xw);                 % Blackman window transform
spec = 20*log10(abs(Xw));     % Spectral magnitude in dB
spec = spec - max(spec);      % Normalize to 0 db max
nfft = zpf*M;
```

---

[3] http://ccrma.stanford.edu/~jos/sasp/Classic_Blackman.html

[4] http://www.octave.org/

[5] http://www.mathworks.com/access/helpdesk/help/toolbox/signal/signal.shtml

```
spec = max(spec,-100*ones(1,nfft)); % clip to -100 dB
fni = [0:1.0/nfft:1-1.0/nfft];   % Normalized frequency axis
subplot(3,1,2); plot(fni,spec,'-'); axis([0,1,-100,10]);
xlabel('Normalized Frequency (cycles per sample))');
ylabel('Magnitude (dB)'); grid; text(-.12,20,'b)');

% Replot interpreting upper bin numbers as frequencies<0:
nh = nfft/2;
specnf = [spec(nh+1:nfft),spec(1:nh)];  % see fftshift()
fninf = fni - 0.5;
subplot(3,1,3);
plot(fninf,specnf,'-'); axis([-0.5,0.5,-100,10]); grid;
xlabel('Normalized Frequency (cycles per sample))');
ylabel('Magnitude (dB)');
text(-.62,20,'c)');
cmd = ['print -deps ', '../eps/blackman.eps'];
disp(cmd); eval(cmd);
disp 'pausing for RETURN (check the plot). . .'; pause
```

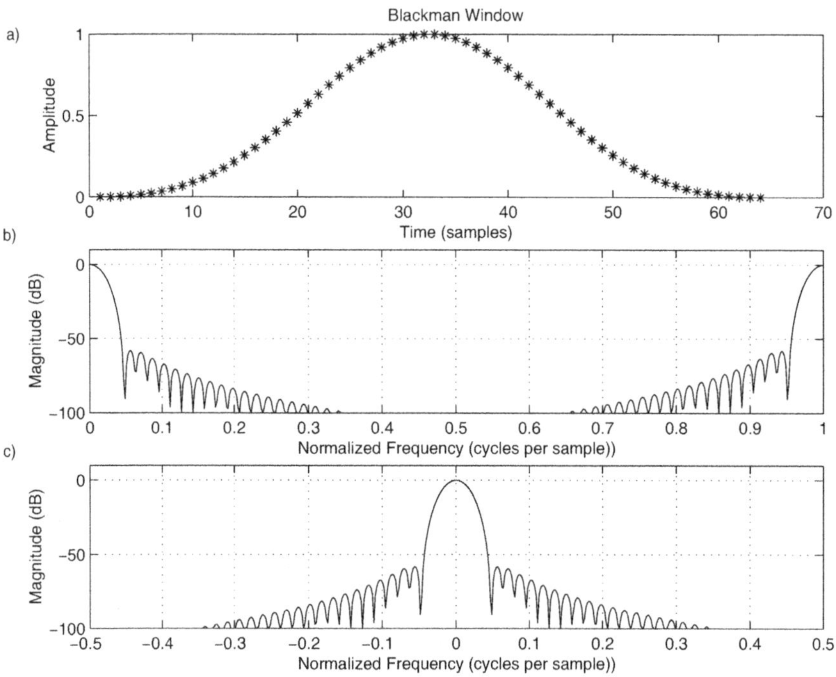

Figure 8.5: The Blackman window: a) window itself in the time domain, b) dB magnitude spectrum plotted over normalized frequencies $[0, 1)$, and c) same thing plotted over $[-0.5, 0.5)$.

Applying the Blackman Window

Now let's apply the Blackman window to the sampled sinusoid and look at the effect on the spectrum analysis:

```
% Windowed, zero-padded data:
n = [0:M-1];            % discrete time axis
f = 0.25 + 0.5/M;       % frequency
xw = [w .* cos(2*pi*n*f),zeros(1,(zpf-1)*M)];

% Smoothed, interpolated spectrum:
X = fft(xw);

% Plot time data:
subplot(2,1,1);
plot(xw);
title('Windowed, Zero-Padded, Sampled Sinusoid');
xlabel('Time (samples)');
ylabel('Amplitude');
text(-50,1,'a)');

% Plot spectral magnitude:
spec = 10*log10(conj(X).*X);  % Spectral magnitude in dB
spec = max(spec,-60*ones(1,nfft)); % clip to -60 dB
subplot(2,1,2);
plot(fninf,fftshift(spec),'-');
axis([-0.5,0.5,-60,40]);
title('Smoothed, Interpolated, Spectral Magnitude (dB)');
xlabel('Normalized Frequency (cycles per sample))');
ylabel('Magnitude (dB)'); grid;
text(-.6,40,'b)');
```

Figure 8.6 plots the zero-padded, Blackman-windowed sinusoid, along with its magnitude spectrum on a dB scale. Note that the first sidelobe (near -40 dB) is nearly 60 dB below the spectral peak (near $+20$ dB). This is why the Blackman window is considered adequate for many audio applications. From the *dual* of the convolution theorem discussed in §7.4.6 on page 155, we know that *windowing in the time domain* corresponds to *smoothing in the frequency domain*. Specifically, the complex spectrum with magnitude displayed in Fig. 8.4b (p. 171) has been *convolved* with the Blackman window transform (dB magnitude shown in Fig. 8.5c). Thus, the Blackman window Fourier transform has been applied as a *smoothing kernel* to the Fourier transform of the rectangularly windowed sinusoid to produce the smoothed result in Fig. 8.6b. This topic is pursued in detail at the outset of Book IV in the music signal processing series [67].

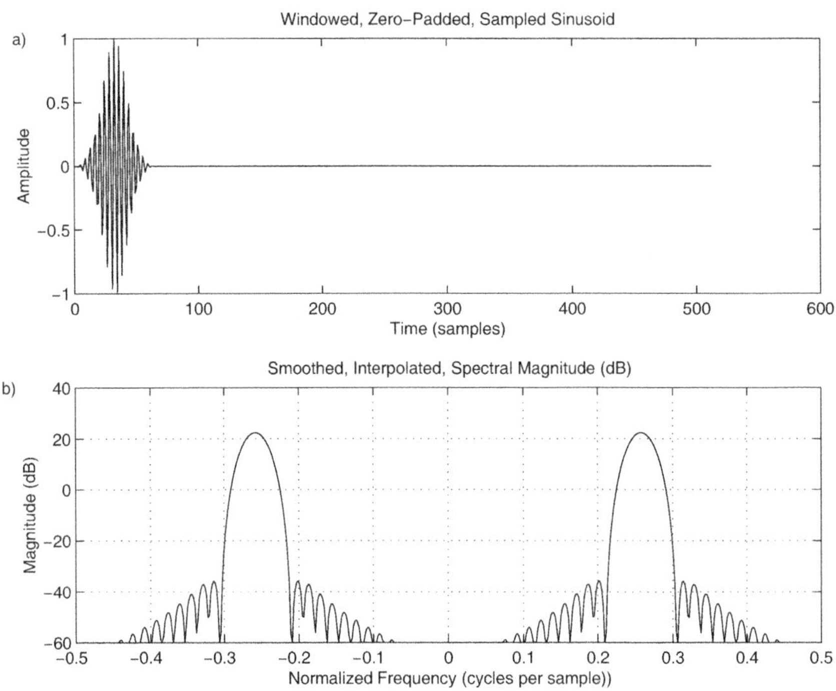

Figure 8.6: Effect of the Blackman window on the sinusoidal data.

8.1.5 Hann-Windowed Complex Sinusoid

In this example, we'll perform spectrum analysis on a *complex* sinusoid having only a single positive frequency. We'll use the *Hann window* (also known as the *Hanning window*) which does not have as much sidelobe suppression as the Blackman window, but its main lobe is narrower. Its sidelobes "roll off" very quickly versus frequency. Compare with the Blackman window results to see if you can see these differences.

The Matlab script for synthesizing and plotting the Hann-windowed sinusoid is given below:

```
% Analysis parameters:
M = 31;          % Window length
N = 64;          % FFT length (zero padding factor near 2)

% Signal parameters:
wxT = 2*pi/4;    % Sinusoid frequency (rad/sample)
A = 1;           % Sinusoid amplitude
phix = 0;        % Sinusoid phase

% Compute the signal x:
n = [0:N-1];     % time indices for sinusoid and FFT
x = A * exp(j*wxT*n+phix); % complex sine [1,j,-1,-j...]

% Compute Hann window:
nm = [0:M-1];    % time indices for window computation
% Hann window = "raised cosine", normalization (1/M)
% chosen to give spectral peak magnitude at 1/2:
w = (1/M) * (cos((pi/M)*(nm-(M-1)/2))).^2;

wzp = [w,zeros(1,N-M)]; % zero-pad out to the length of x
xw = x .* wzp;          % apply the window w to signal x

figure(1);
subplot(1,1,1);

% Display real part of windowed signal and Hann window
plot(n,wzp,'-k'); hold on; plot(n,real(xw),'*k'); hold off;
title(['Hann Window and Windowed, Zero-Padded, ',...
       'Sinusoid (Real Part)']);
xlabel('Time (samples)'); ylabel('Amplitude');
```

The resulting plot of the Hann window and its use on sinusoidal data are shown in Fig. 8.7.

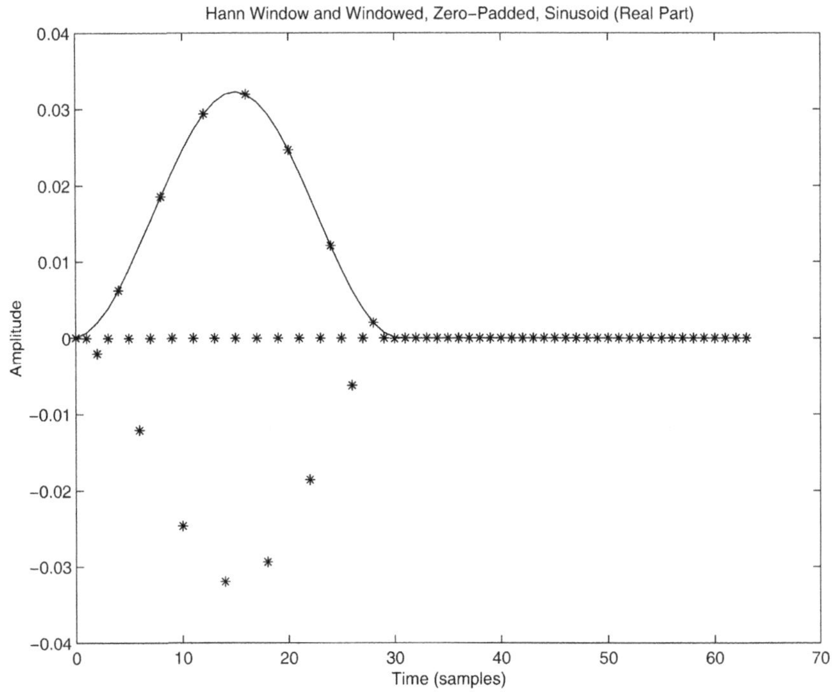

Figure 8.7: A length 31 Hann window ("raised cosine") overlaid with the real part of the Hann-windowed complex sinusoid. Zero-padding is also shown. The sampled sinusoid is plotted using '*' with no connecting interpolation lines. You must now imagine the continuous real sinusoid (windowed) threading through the asterisks.

Hann Window Spectrum Analysis Results

Finally, the Matlab for computing the DFT of the Hann-windowed complex sinusoid and plotting the results is listed below. To help see the full spectrum, we also compute a heavily interpolated spectrum (via zero padding as before) which we'll draw using solid lines.

```
% Compute the spectrum and its alternative forms:
Xw = fft(xw);                % FFT of windowed data
fn = [0:1.0/N:1-1.0/N];      % Normalized frequency axis
spec = 20*log10(abs(Xw));    % Spectral magnitude in dB
% Since the nulls can go to minus infinity, clip at -100 dB:
spec = max(spec,-100*ones(1,length(spec)));
phs = angle(Xw);             % Spectral phase in radians
phsu = unwrap(phs);          % Unwrapped spectral phase

% Compute heavily interpolated versions for comparison:
Nzp = 16;                     % Zero-padding factor
Nfft = N*Nzp;                 % Increased FFT size
xwi = [xw,zeros(1,Nfft-N)];   % New zero-padded FFT buffer
Xwi = fft(xwi);               % Compute interpolated spectrum
fni = [0:1.0/Nfft:1.0-1.0/Nfft]; % Normalized freq axis
speci = 20*log10(abs(Xwi));   % Interpolated spec mag (dB)
speci = max(speci,-100*ones(1,length(speci))); % clip
phsi = angle(Xwi);            % Interpolated phase
phsiu = unwrap(phsi);         % Unwrapped interpolated phase

figure(1);
subplot(2,1,1);

plot(fn,abs(Xw),'*k'); hold on;
plot(fni,abs(Xwi),'-k'); hold off;
title('Spectral Magnitude');
xlabel('Normalized Frequency (cycles per sample))');
ylabel('Amplitude (linear)');

subplot(2,1,2);

% Same thing on a dB scale
plot(fn,spec,'*k'); hold on; plot(fni,speci,'-k'); hold off;
title('Spectral Magnitude (dB)');
xlabel('Normalized Frequency (cycles per sample))');
ylabel('Magnitude (dB)');

cmd = ['print -deps ', 'specmag.eps']; disp(cmd); eval(cmd);
disp 'pausing for RETURN (check the plot). . .'; pause
```

```
figure(1);
subplot(2,1,1);
plot(fn,phs,'*k'); hold on; plot(fni,phsi,'-k'); hold off;
title('Spectral Phase');
xlabel('Normalized Frequency (cycles per sample))');
ylabel('Phase (rad)'); grid;
subplot(2,1,2);
plot(fn,phsu,'*k'); hold on; plot(fni,phsiu,'-k'); hold off;
title('Unwrapped Spectral Phase');
xlabel('Normalized Frequency (cycles per sample))');
ylabel('Phase (rad)'); grid;
cmd = ['print -deps ', 'specphs.eps']; disp(cmd); eval(cmd);
```

Figure 8.8 shows the spectral magnitude and Fig. 8.9 the spectral phase.

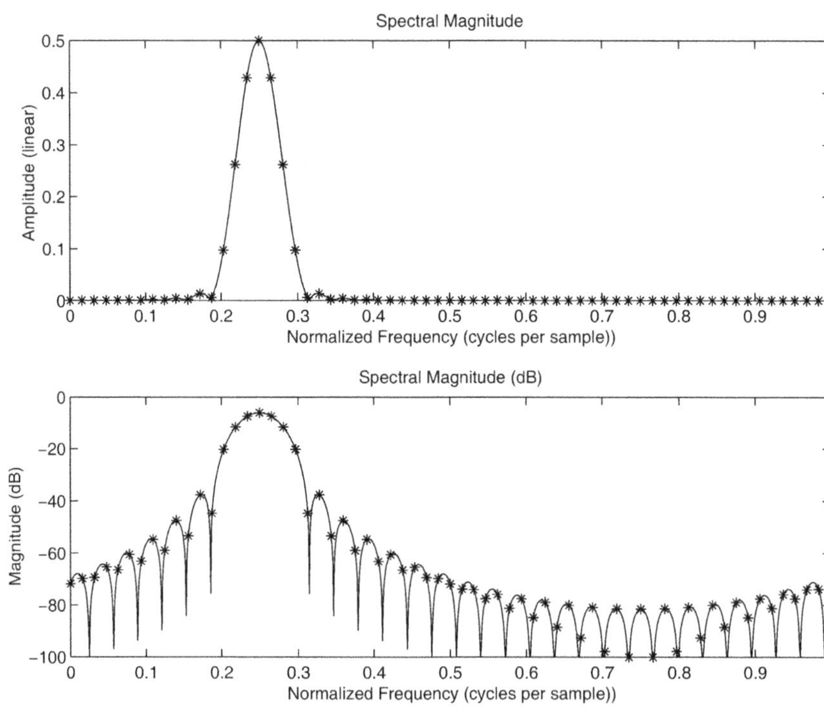

Figure 8.8: Spectral magnitude on linear (top) and dB (bottom) scales.

There are no negative-frequency components in Fig. 8.8 because we are analyzing a complex sinusoid $x = [1, j, -1, -j, 1, j, \dots]$, which has frequency $f_s/4$ only, with no component at $-f_s/4$.

Notice how difficult it would be to correctly interpret the shape of the "sidelobes" without zero padding. The asterisks correspond to a zero-padding factor of 2, already twice as much as needed to preserve all spectral information faithfully, but not enough to clearly outline the sidelobes in a spectral magnitude plot.

Spectral Phase

As for the phase of the spectrum, what do we expect? We have chosen the sinusoid phase offset to be zero. The window is causal and symmetric about its middle. Therefore, we expect a linear phase term with slope $-(M-1)/2 = -15$ samples (as discussed in connection with the shift theorem in §7.4.4 on page 149). Also, the window transform has sidelobes which cause a phase of π radians to switch in and out. Thus, we expect to see samples of a straight line (with slope -15 samples) across the main lobe of the window transform, together with a switching offset by π in every other sidelobe away from the main lobe, starting with the immediately adjacent sidelobes.

In Fig. 8.9(a), we can see the negatively sloped line across the main lobe of the window transform, but the sidelobes are hard to follow. Even the unwrapped phase in Fig. 8.9(b) is not as clear as it could be. This is because a phase jump of π radians and $-\pi$ radians are equally valid, as is any odd multiple of π radians. In the case of the unwrapped phase, all phase jumps are by $+\pi$ starting near frequency 0.3. Figure 8.9(c) shows what could be considered the "canonical" unwrapped phase for this example: We see a linear phase segment across the main lobe as before, and outside the main lobe, we have a continuation of that linear phase across all of the positive sidelobes, and only a π-radian deviation from that linear phase across the negative sidelobes. In other words, we see a straight linear phase at the desired slope interrupted by temporary jumps of π radians. To obtain unwrapped phase of this type, the `unwrap` function needs to alternate the sign of successive phase-jumps by π radians; this could be implemented, for example, by detecting jumps-by-π to within some numerical tolerance and using a bit of state to enforce alternation of $+\pi$ with $-\pi$.

To convert the expected phase slope from -15 "radians per (rad/sec)" to "radians per cycle-per-sample," we need to multiply by "radians per cycle," or 2π. Thus, in Fig. 8.9(c), we expect a slope of -94.2 radians per unit normalized frequency, or -9.42 radians per 0.1 cycles-per-sample, and this looks about right, judging from the plot.

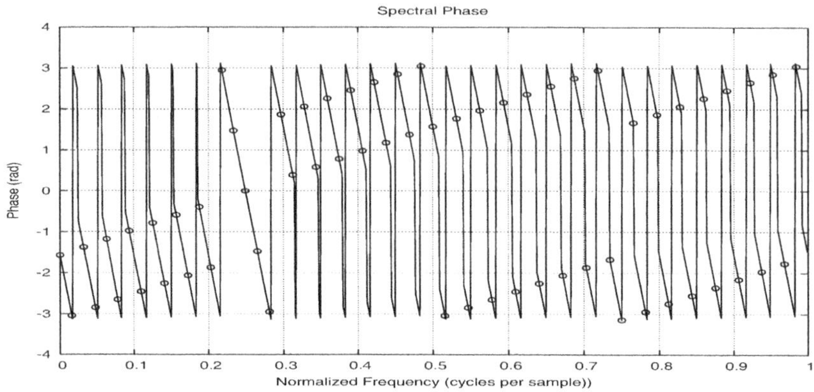

(a) Raw spectral phase and its interpolation

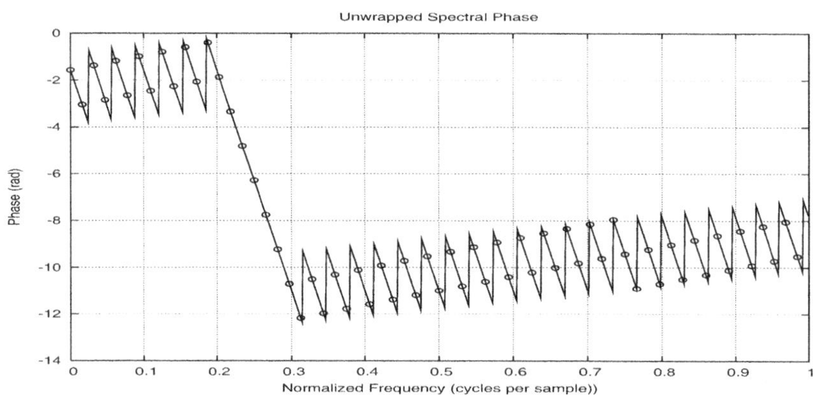

(b) Unwrapped spectral phase and its interpolation

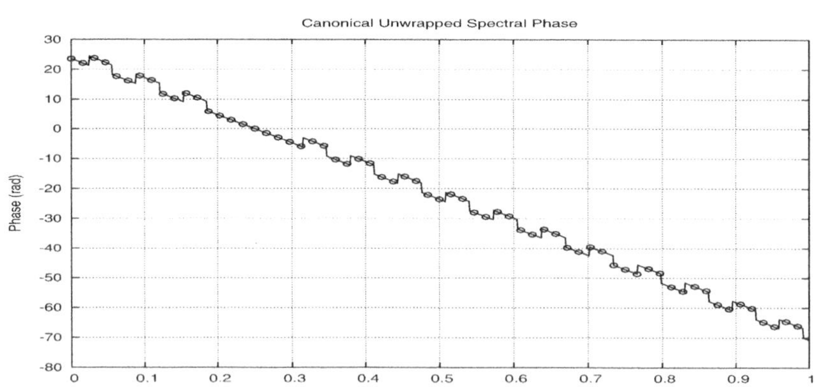

(c) Canonically unwrapped spectral phase and its interpolation

Figure 8.9: Spectral phase and two different phase unwrappings.

8.2 Spectrograms

The *spectrogram* is a basic tool in audio spectral analysis and other fields. It has been applied extensively in speech analysis [16]. The spectrogram can be defined as an intensity plot (usually on a log scale, such as dB) of the Short-Time Fourier Transform (STFT) magnitude. The STFT is simply a sequence of FFTs of windowed data segments, where the windows are usually allowed to overlap in time, typically by 25–50% [3]. It is an important representation of audio data because human hearing is based on a kind of real-time spectrogram encoded by the cochlea of the inner ear [47]. The spectrogram has been used extensively in the field of computer music as a guide during the development of sound synthesis algorithms. When working with an appropriate synthesis model, matching the spectrogram often corresponds to matching the sound extremely well. In fact, *spectral modeling synthesis* (SMS) is based on synthesizing the short-time spectrum directly by some means [83].

Spectrogram of Speech

An example spectrogram for recorded *speech data* is shown in Fig. 8.10. It was generated using the Matlab code displayed in Fig. 8.11. The function `spectrogram` is listed in §I.5 on page 287. The spectrogram is computed as a sequence of FFTs of windowed data segments. The spectrogram is plotted by `spectrogram` using `imagesc`.

In this example, the Hamming window length was chosen to be 20 ms, as is typical in speech analysis. This is short enough so that any single 20 ms frame will typically contain data from only one phoneme,[6] yet long enough that it will include at least two periods of the fundamental frequency during voiced speech, assuming the lowest voiced pitch to be around 100 Hz.

More generally, for speech and the singing voice (and any periodic tone), the STFT analysis parameters are chosen to trade off among the following conflicting criteria:

1. The *harmonics* should be resolved.

2. *Pitch* and *formant* variations should be closely followed.

The *formants* in speech are the resonances in the vocal tract. They appear as dark groups of harmonics in Fig. 8.10. The first two formants largely determine the "vowel" in voiced speech. In telephone speech, nominally between 200 and 3200 Hz, only three or four formants are usually present in the band.

[6] A *phoneme* is a single elementary sound in speech, such as a single vowel sound like the 'a' in "bat" or the 's' in "sit".

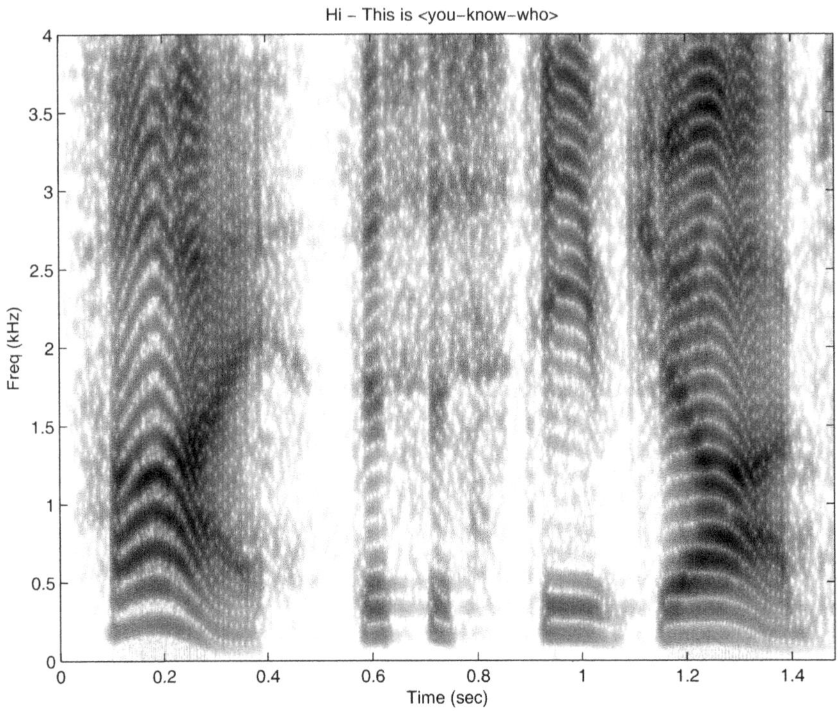

Figure 8.10: Classic spectrogram of a speech sample.

```
[y,fs,bits] = wavread('SpeechSample.wav');
soundsc(y,fs); % Let's hear it
% for classic look:
colormap('gray'); map = colormap; imap = flipud(map);
M = round(0.02*fs);  % 20 ms window is typical
N = 2^nextpow2(4*M); % zero padding for interpolation
w = 0.54 - 0.46 * cos(2*pi*[0:M-1]/(M-1)); % w = hamming(M);
colormap(imap); % Octave wants it here
spectrogram(y,N,fs,w,-M/8,1,60);
colormap(imap); % Matlab wants it here
title('Hi - This is <you-know-who> ');
ylim([0,(fs/2)/1000]); % don't plot neg. frequencies
```

Figure 8.11: Matlab for computing a speech spectrogram.

8.3 Filters and Convolution

A reason for the importance of convolution (defined in §7.2.4 on page 121) is that *every linear time-invariant system*[7] *can be represented by a convolution.* Thus, in the convolution equation

$$y = h * x \tag{8.1}$$

we may interpret x as the *input* signal to a filter, y as the *output* signal, and h as the *digital filter*, as shown in Fig. 8.12.

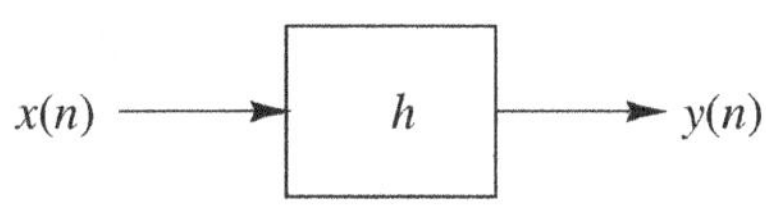

Figure 8.12: The filter interpretation of convolution.

The *impulse* or "unit pulse" signal is defined by

$$\delta(n) \triangleq \begin{cases} 1, & n = 0 \\ 0, & n \neq 0. \end{cases}$$

For example, for sequences of length $N = 4$, $\delta = [1, 0, 0, 0]$.

The impulse signal is the *identity element* under convolution, since

$$(x * \delta)_n \triangleq \sum_{m=0}^{N-1} x(m)\delta(n - m) = x(n).$$

If we set $x = \delta$ in Eq. (8.1) above, we get

$$y = h * \delta = h.$$

Thus, h, which we introduced as the convolution representation of a filter, has been shown to be more specifically the *impulse response* of the filter.

It turns out in general that every linear time-invariant (LTI) system (filter) is completely described by its impulse response [65]. No matter what the LTI system is, we can feed it an impulse, record what comes out, call it $h(n)$, and implement the system by convolving the input signal x with the impulse response h. In other words, every LTI system has a convolution representation in terms of its impulse response.

[7]Linearity and time invariance are introduced in the second book of this series [65].

8.3.1 Frequency Response

Definition: The *frequency response* of an LTI filter may be defined as the Fourier transform of its impulse response. In particular, for finite, discrete-time signals $h \in \mathbf{C}^N$, the sampled frequency response may be defined as

$$H(\omega_k) \triangleq \text{DFT}_k(h).$$

The complete (continuous) frequency response is defined using the DTFT (see §B.1 on page 208), *i.e.*,

$$H(\omega) \triangleq \text{DTFT}_\omega(\text{ZeroPad}_\infty(h)) \triangleq \sum_{n=0}^{N-1} h(n)e^{-j\omega n}$$

where the summation limits are truncated to $[0, N-1]$ because $h(n)$ is zero for $n < 0$ and $n > N-1$. Thus, the DTFT can be obtained from the DFT by simply replacing ω_k by ω, which corresponds to infinite zero-padding in the time domain. Recall from §7.2.10 on page 134 that zero-padding in the time domain gives ideal interpolation of the frequency-domain samples $H(\omega_k)$ (assuming the original DFT included all nonzero samples of h).

8.3.2 Amplitude Response

Definition: The *amplitude response* of a filter is defined as the *magnitude* of the frequency response

$$G(k) \triangleq |H(\omega_k)|.$$

From the convolution theorem, we can see that the amplitude response $G(k)$ is the *gain* of the filter at frequency ω_k, since

$$|Y(\omega_k)| = |H(\omega_k)X(\omega_k)| = G(k)\,|X(\omega_k)|,$$

where $X(\omega_k)$ is the kth sample of the DFT of the input signal $x(n)$, and Y is the DFT of the output signal y.

8.3.3 Phase Response

Definition: The *phase response* of a filter is defined as the *phase* of its frequency response:

$$\Theta(k) \triangleq \angle H(\omega_k)$$

From the convolution theorem, we can see that the phase response $\Theta(k)$ is the phase-shift added by the filter to an input sinusoidal component at frequency ω_k, since

$$\angle Y(\omega_k) = \angle[H(\omega_k)X(\omega_k)] = \angle H(\omega_k) + \angle X(\omega_k) = \Theta(k) + \angle X(\omega_k).$$

The topics touched upon in this section are developed more fully in the next book [65] in the music signal processing series mentioned in the preface.

8.4 Correlation Analysis

The *correlation operator* (defined in §7.2.5 on page 129) plays a major role in statistical signal processing. For a proper development, see, *e.g.*, [25, 31, 62]. This section introduces only some of the most basic elements of statistical signal processing in a simplified manner, with emphasis on illustrating applications of the DFT.

8.4.1 Cross-Correlation

Definition: The *circular cross-correlation* of two signals x and y in $\mathbf{C}^N$ may be defined by

$$\boxed{\hat{r}_{xy}(l) \stackrel{\Delta}{=} \frac{1}{N}(x \star y)(l) \stackrel{\Delta}{=} \frac{1}{N}\sum_{n=0}^{N-1} \overline{x(n)}y(n+l),\; l = 0, 1, 2, \ldots, N-1.}$$

(Note that the "lag" l is an integer variable, not the constant 1.) The *DFT correlation operator* '$\star$' was first defined in §7.2.5 on page 129.

The term "cross-correlation" comes from *statistics*, and what we have defined here is more properly called a "sample cross-correlation." That is, $\hat{r}_{xy}(l)$ is an *estimator*[8] of the true cross-correlation $r_{xy}(l)$ which is an assumed statistical property of the signal itself. This definition of a sample cross-correlation is only valid for *stationary* stochastic processes, *e.g.*, "steady noises" that sound unchanged over time. The statistics of a stationary stochastic process are by definition *time invariant*, thereby allowing *time-averages* to be used for estimating statistics such as cross-correlations. For brevity below, we will typically *not* include "sample" qualifier, because all computational methods discussed will be sample-based methods intended for use on stationary data segments.

[8]In signal processing, a "hat" often denotes an *estimated* quantity. Thus, $\hat{r}_{xy}(l)$ is an estimate of $r_{xy}(l)$.

The DFT of the cross-correlation may be called the *cross-spectral density*, or "cross-power spectrum," or even simply "cross-spectrum":

$$\hat{R}_{xy}(\omega_k) \triangleq \text{DFT}_k(\hat{r}_{xy}) = \frac{\overline{X(\omega_k)}Y(\omega_k)}{N}$$

The last equality above follows from the correlation theorem (§7.4.7 on page 155).

8.4.2 Unbiased Cross-Correlation

Recall that the cross-correlation operator is *cyclic* (circular) since $n+l$ is interpreted modulo N. In practice, we are normally interested in estimating the *acyclic* cross-correlation between two signals. For this (more realistic) case, we may define instead the *unbiased cross-correlation*

$$\boxed{\hat{r}^u_{xy}(l) \triangleq \frac{1}{N-l}\sum_{n=0}^{N-1-l}\overline{x(n)}y(n+l), \quad l=0,1,2,\ldots,L-1}$$

where we choose $L \ll N$ (*e.g.*, $L \approx \sqrt{N}$) in order to have enough lagged products $\overline{x(n)}y(n+l)$ at the highest lag $L-1$ so that a reasonably accurate average is obtained. Note that the summation stops at $n = N-l-1$ to avoid cyclic wrap-around of n modulo N. The term "unbiased" refers to the fact that the expected value[9] [31] of $\hat{r}^u_{xy}(l)$ is the true cross-correlation $r_{xy}(l)$ of x and y (assumed to be samples from stationary stochastic processes).

An unbiased acyclic cross-correlation may be computed faster via DFT (FFT) methods using zero padding:

$$\boxed{\hat{r}^u_{xy}(l) = \frac{1}{N-l}\text{IDFT}_l(\overline{X}\cdot Y), \quad l=0,1,2,\ldots,L-1}$$

where

$$\begin{aligned} X &= \text{DFT}[\text{CausalZeroPad}_{N+L-1}(x)] \\ Y &= \text{DFT}[\text{CausalZeroPad}_{N+L-1}(y)]. \end{aligned}$$

Note that x and y belong to $\mathbf{C}^N$ while X and Y belong to $\mathbf{C}^{N+L-1}$. The zero-padding may be causal (as defined in §7.2.8 on page 133) because the signals are assumed to be be stationary, in which case all signal statistics

[9] For present purposes, the expected value $r_{xy}(l)$ may be found by averaging an infinite number of cross-correlations $\hat{r}^u_{xy}(l)$ computed using different segments of x and y. Both x and y must be infinitely long, of course, and all stationary processes *are* infinitely long. Otherwise, their statistics could not be time invariant.

are time-invariant. As usual when embedding acyclic correlation (or convolution) within the cyclic variant given by the DFT, sufficient zero-padding is provided so that only zeros are "time aliased" (wrapped around in time) by modulo indexing.

Cross-correlation is used extensively in audio signal processing for applications such as *time scale modification*, *pitch shifting*, *click removal*, and many others.

8.4.3 Autocorrelation

The cross-correlation of a signal with itself gives its *autocorrelation*:

$$\boxed{\hat{r}_x(l) \triangleq \frac{1}{N}(x \star x)(l) \triangleq \frac{1}{N}\sum_{n=0}^{N-1} \overline{x(n)}x(n+l)}$$

The autocorrelation function is Hermitian:

$$\hat{r}_x(-l) = \overline{\hat{r}_x(l)}$$

When x is real, its autocorrelation is *real* and *even* (symmetric about lag zero).

The unbiased cross-correlation similarly reduces to an unbiased autocorrelation when $x \equiv y$:

$$\boxed{\hat{r}_x^u(l) \triangleq \frac{1}{N-l}\sum_{n=0}^{N-1-l} \overline{x(n)}x(n+l), \quad l = 0, 1, 2, \dots, L-1} \tag{8.2}$$

The DFT of the true autocorrelation function $r_x(n) \in \mathbf{R}^N$ is the (sampled) *power spectral density* (PSD), or *power spectrum*, and may be denoted

$$R_x(\omega_k) \triangleq \text{DFT}_k(r_x).$$

The complete (not sampled) PSD is $R_x(\omega) \triangleq \text{DTFT}_k(r_x)$, where the DTFT is defined in Appendix B (it's just an infinitely long DFT). The DFT of $\hat{r}_x$ thus provides a sample-based *estimate* of the PSD:[10]

$$\hat{R}_x(\omega_k) = \text{DFT}_k(\hat{r}_x) = \frac{|X(\omega_k)|^2}{N}$$

[10]To clarify, we are using the word "sample" with two different meanings. In addition to the usual meaning wherein a continuous time or frequency axis is made discrete, a statistical "sample" refers to a set of observations from some presumed random process. Estimated statistics based on such a statistical sample are then called "sample statistics", such as the sample mean, sample variance, and so on.

We could call $\hat{R}_x(\omega_k)$ a "sampled sample power spectral density".

At lag zero, the autocorrelation function reduces to the *average power* (mean square) which we defined in §5.8 on page 74:

$$\hat{r}_x(0) \stackrel{\Delta}{=} \frac{1}{N} \sum_{m=0}^{N-1} |x(m)|^2$$

Replacing "correlation" with "covariance" in the above definitions gives corresponding zero-mean versions. For example, we may define the *sample circular cross-covariance* as

$$\boxed{\hat{c}_{xy}(n) \stackrel{\Delta}{=} \frac{1}{N} \sum_{m=0}^{N-1} \overline{[x(m) - \mu_x]}[y(m+n) - \mu_y].}$$

where μ_x and μ_y denote the means of x and y, respectively. We also have that $\hat{c}_x(0)$ equals the sample *variance* of the signal x:

$$\hat{c}_x(0) \stackrel{\Delta}{=} \frac{1}{N} \sum_{m=0}^{N-1} |x(m) - \mu_x|^2 \stackrel{\Delta}{=} \hat{\sigma}_x^2$$

8.4.4 Matched Filtering

The cross-correlation function is used extensively in *pattern recognition* and *signal detection*. We know from Chapter 5 that projecting one signal onto another is a means of measuring how much of the second signal is present in the first. This can be used to "detect" the presence of known signals as components of more complicated signals. As a simple example, suppose we record $x(n)$ which we think consists of a signal $s(n)$ that we are looking for plus some additive measurement noise $e(n)$. That is, we assume the signal model $x(n) = s(n) + e(n)$. Then the projection of x onto s is (recalling §5.9.9 on page 86)

$$\mathbf{P}_s(x) \stackrel{\Delta}{=} \frac{\langle x, s\rangle}{\|s\|^2} s = \frac{\langle s+e, s\rangle}{\|s\|^2} s = s + \frac{\langle e, s\rangle}{\|s\|^2} s = s + \frac{N}{\|s\|^2} \hat{r}_{se}(0) s \approx s$$

since the projection of random, zero-mean noise e onto s is small with probability one. Another term for this process is *matched filtering*. The impulse response of the "matched filter" for a real signal s is given by $\text{FLIP}(s)$.[11] By time-reversing s, we transform the convolution implemented by filtering into a sliding cross-correlation operation between the input signal x and

[11]See Eq. (7.1) on page 118 for a definition of $\text{FLIP}()$.

the sought signal s. (For complex known signals s, the matched filter is $\text{FLIP}(\overline{s})$.) We detect occurrences of s in x by detecting peaks in $\hat{r}_{sx}(l)$.

In the same way that FFT convolution is faster than direct convolution (see Table 7.1 on page 154), cross-correlation and matched filtering are generally carried out most efficiently using an FFT algorithm (Appendix A).

8.4.5 FIR System Identification

Estimating an impulse response from input-output measurements is called *system identification*, and a large literature exists on this topic (*e.g.*, [37]).

Cross-correlation can be used to compute the impulse response $h(n)$ of a filter from the cross-correlation of its input and output signals $x(n)$ and $y = h * x$, respectively. To see this, note that, by the correlation theorem,

$$x \star y \longleftrightarrow \overline{X} \cdot Y = \overline{X} \cdot (H \cdot X) = H \cdot |X|^2 .$$

Therefore, the frequency response equals the input-output cross-spectrum divided by the input power spectrum:

$$H = \frac{\overline{X} \cdot Y}{|X|^2} = \frac{\hat{R}_{xy}}{\hat{R}_{xx}}$$

where multiplication and division of spectra are defined pointwise, *i.e.*, $H(\omega_k) = \overline{X(\omega_k)} \cdot Y(\omega_k)/|X(\omega_k)|^2$. A Matlab program illustrating these relationships is listed in Fig. 8.13.

8.5 Power Spectral Density Estimation

Welch's method [82] (or the *periodogram method* [18]) for estimating power spectral densities (PSD) is carried out by dividing the time signal into successive blocks, and *averaging squared-magnitude DFTs of the signal blocks.* Let $x_m(n) = x(n + mN)$, $n = 0, 1, \ldots, N - 1$, denote the mth block of the signal $x \in \mathbf{C}^{MN}$, with M denoting the number of blocks. Then the Welch PSD estimate is given by

$$\hat{R}_x(\omega_k) = \frac{1}{M} \sum_{m=0}^{M-1} |DFT_k(x_m)|^2 \triangleq \left\{|X_m(\omega_k)^2|\right\}_m \tag{8.3}$$

where "$\{\cdot\}_m$" denotes *time averaging* across blocks (or "frames") of data indexed by m. The function `pwelch` implements Welch's method in Octave (Octave-Forge collection) and Matlab (Signal Processing Toolbox).

```
% sidex.m - Demonstration of the use of FFT cross-
% correlation to compute the impulse response
% of a filter given its input and output.
% This is called "FIR system identification".

Nx = 32; % input signal length
Nh = 10; % filter length Ny = Nx+Nh-1;
% max output signal length
% FFT size to accommodate cross-correlation:
Nfft = 2^nextpow2(Nx+Nh-1); % FFT wants power of 2

x = rand(1,Nx); % input signal = noise
%x = 1:Nx;  % input signal = ramp
h = [1:Nh];  % the filter
xzp = [x,zeros(1,Nfft-Nx)]; % zero-padded input
yzp = filter(h,1,xzp); % apply the filter
X = fft(xzp);   % input spectrum
Y = fft(yzp);   % output spectrum
Rxx = conj(X) .* X; % energy spectrum of x
Rxy = conj(X) .* Y; % cross-energy spectrum
Hxy = Rxy ./ Rxx;   % should be the freq. response
hxy = ifft(Hxy);    % should be the imp. response

hxy(1:Nh)      % print estimated impulse response
freqz(hxy,1,Nfft);  % plot estimated freq response

err = norm(hxy - [h,zeros(1,Nfft-Nh)])/norm(h);
disp(sprintf(['Impulse Response Error = ',...
'%0.14f%%'],100*err));

err = norm(Hxy-fft([h,zeros(1,Nfft-Nh)]))/norm(h);
disp(sprintf(['Frequency Response Error = ',...
'%0.14f%%'],100*err));
```

Figure 8.13: FIR system identification example in matlab.

Recall that $|X_m|^2 \leftrightarrow x \star x$ which is *circular* (cyclic) autocorrelation. To obtain an *acyclic* autocorrelation instead, we may use *zero padding* in the time domain, as described in §8.4.2 on page 188. That is, we can replace x_m above by $\text{CausalZeroPad}_{2N-1}(x_m) = [x_m, 0, \ldots, 0]$.[12] Although this fixes the "wrap-around problem", the estimator is still *biased* because its expected value is the true autocorrelation $r_x(l)$ weighted by $N - |l|$. This bias is equivalent to multiplying the correlation in the "lag domain" by a *triangular window* (also called a "Bartlett window"). The bias can be removed by simply dividing it out, as in Eq. (8.2), but it is common to retain the Bartlett weighting since it merely corresponds to *smoothing* the power spectrum (or cross-spectrum) with a sinc^2 kernel;[13] it also down-weights the less reliable large-lag estimates, weighting each lag by the number of lagged products that were summed.

Since $|X_m(\omega_k)|^2 = N \cdot \text{DFT}_k(\hat{r}_{x_m})$, and since the DFT is a *linear operator* (§7.4.1), averaging magnitude-squared DFTs $|X_m(\omega_k)|^2$ is *equivalent*, in principle, to estimating block autocorrelations $\hat{r}_{x_m}$, averaging them, and taking a DFT of the average. However, this would normally be slower.

We return to power spectral density estimation in Book IV [67] of the music signal processing series.

8.6 Coherence Function

A function related to cross-correlation is the *coherence function*, defined in terms of power spectral densities and the cross-spectral density by

$$C_{xy}(\omega) \triangleq \frac{|R_{xy}(\omega)|^2}{R_x(\omega)R_y(\omega)}.$$

In practice, these quantities can be estimated by *time-averaging* $\overline{X(\omega_k)}Y(\omega_k)$, $|X(\omega_k)|^2$, and $|Y(\omega_k)|^2$ over successive signal blocks. Let $\{\cdot\}_m$ denote time averaging across frames as in Eq. (8.3) above. Then an estimate of the coherence, the *sample coherence function* $\hat{C}_{xy}(\omega_k)$, may be defined by

$$\hat{C}_{xy}(\omega_k) \triangleq \frac{\left|\left\{\overline{X_m(\omega_k)}Y_m(\omega_k)\right\}_m\right|^2}{\left\{|X_m(\omega_k)|^2\right\}_m \cdot \left\{|Y_m(\omega_k)|^2\right\}_m}.$$

---

[12] Since phase information is discarded ($x_m \star x_m \leftrightarrow |X_m(\omega_k)|^2$), the zero-padding can go before or after x_m, or both, without affecting the results.

[13] By the convolution theorem dual, windowing in the time domain is convolution (smoothing) in the frequency domain (§7.4.6). Since a triangle is the convolution of a rectangle with itself, its transform is sinc^2 in the continuous-time case (cf. Appendix D). In the discrete-time case, it is proportional to $\text{Alias}_{2\pi/T}(\text{sinc}^2)$.

Note that the averaging in the numerator occurs before the absolute value is taken.

The coherence $C_{xy}(\omega)$ is a real function between zero and one which gives a *measure of correlation* between x and y at each frequency ω. For example, imagine that y is produced from x via an LTI filtering operation:

$$y = h * x \implies Y(\omega_k) = H(\omega_k)X(\omega_k)$$

Then the magnitude-normalized cross-spectrum in each frame is

$$\begin{aligned}\hat{A}_{x_m y_m}(\omega_k) &\triangleq \frac{\overline{X_m(\omega_k)}Y_m(\omega_k)}{|X_m(\omega_k)| \cdot |Y_m(\omega_k)|} = \frac{\overline{X_m(\omega_k)}H(\omega_k)X_m(\omega_k)}{|X_m(\omega_k)| \cdot |H(\omega_k)X_m(\omega_k)|} \\ &= \frac{|X_m(\omega_k)|^2 H(\omega_k)}{|X_m(\omega_k)|^2 |H(\omega_k)|} = \frac{H(\omega_k)}{|H(\omega_k)|}\end{aligned}$$

so that the coherence function becomes

$$\left|\hat{C}_{xy}(\omega_k)\right|^2 = \left|\frac{H(\omega_k)}{|H(\omega_k)|}\right|^2 = 1.$$

On the other hand, when x and y are uncorrelated (*e.g.*, y is a noise process not derived from x), the sample coherence converges to *zero* at all frequencies, as the number of blocks in the average goes to infinity.

A common use for the coherence function is in the validation of input/output data collected in an acoustics experiment for purposes of *system identification*. For example, $x(n)$ might be a known signal which is input to an unknown system, such as a reverberant room, say, and $y(n)$ is the recorded response of the room. Ideally, the coherence should be 1 at all frequencies. However, if the microphone is situated at a *null* in the room response for some frequency, it may record mostly noise at that frequency. This is indicated in the measured coherence by a significant dip below 1. An example is shown in Book III [66] for the case of a measured guitar-bridge admittance.[14] A more elementary example is given in the next section.

[14]`http://ccrma.stanford.edu/~jos/pasp/Guitar_Bridge.html#fig:lguitarcoh`

Coherence Function in Matlab

In Matlab and Octave, `cohere(x,y,M)` computes the coherence function C_{xy} using successive DFTs of length M with a Hanning window and 50% overlap. (The window and overlap can be controlled via additional optional arguments.) The matlab listing in Fig. 8.14 illustrates `cohere` on a simple example. Figure 8.15 shows a plot of `cxyM` for this example. We see a coherence peak at frequency 0.25 cycles/sample, as expected, but there are also two rather large coherence samples on either side of the main peak. These are expected as well, since the true cross-spectrum for this case is a critically sampled Hanning window transform. (A window transform is critically sampled whenever the window length equals the DFT length.)

```
% Illustrate estimation of coherence function 'cohere'
% in the Matlab Signal Processing Toolbox
% or Octave with Octave Forge:
N = 1024;               % number of samples
x=randn(1,N);           % Gaussian noise
y=randn(1,N);           % Uncorrelated noise
f0 = 1/4;               % Frequency of high coherence
nT = [0:N-1];           % Time axis
w0 = 2*pi*f0;
x = x + cos(w0*nT); % Let something be correlated
p = 2*pi*rand(1,1); % Phase is irrelevant
y = y + cos(w0*nT+p);
M = round(sqrt(N)); % Typical window length
[cxyM,w] = cohere(x,y,M); % Do the work
figure(1); clf;
stem(w/2,cxyM,'*'); % w goes from 0 to 1 (odd convention)
legend('');             % needed in Octave
grid on;
ylabel('Coherence');
xlabel('Normalized Frequency (cycles/sample)');
axis([0 1/2 0 1]);
replot; % Needed in Octave
saveplot('../eps/coherex.eps'); % compatibility utility
```

Figure 8.14: Coherence measurement example in matlab.

Note that more than one frame must be averaged to obtain a coherence less than one. For example, changing the `cohere` call in the above example to "`cxyN = cohere(x,y,N);`" produces all ones in `cxyN`, because no averaging is performed.

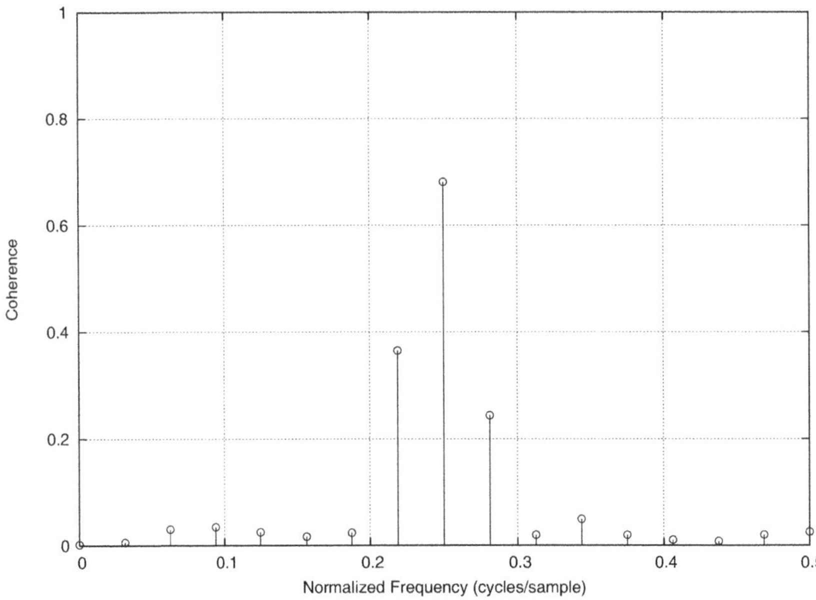

Figure 8.15: Sample coherence function.

Recommended Further Reading

We are now finished developing the mathematics of the DFT and a first look at some of its applications. The sequel consists of appendices which fill in more elementary background and supplement the prior development with related new topics, such as the Fourier transform and FFT algorithm.

For further study, one may, of course, continue on to Book II (**Introduction to Digital Filter Theory** [65]) in the music signal processing series (mentioned in the preface). Alternatively and in addition, the references cited in the bibliography can provide further guidance.

Appendix A

Fast Fourier Transform (FFT) Algorithms

The term *fast Fourier transform* (FFT) refers to an efficient implementation of the discrete Fourier transform (DFT) for highly composite[1] transform lengths N. When computing the DFT as a set of N inner products of length N each, the computational complexity is $\mathcal{O}(N^2)$. When N is an integer power of 2, a Cooley-Tukey FFT algorithm delivers complexity $\mathcal{O}(N \lg N)$, where $\lg N$ denotes the log-base-2 of N, and $\mathcal{O}(x)$ means "on the order of x". Such FFT algorithms were evidently first used by Gauss in 1805 [28] and rediscovered in the 1960s by Cooley and Tukey [14].

In this appendix, a brief introduction is given for various FFT algorithms. A tutorial review (1990) is given in [20]. Additionally, there are some excellent FFT "home pages":

- `http://en.wikipedia.org/wiki/Category:FFT_algorithms`
- `http://faculty.prairiestate.edu/skifowit/fft/`

Pointers to FFT software are given in §A.7 on page 205.

[1]In this context, "highly composite" means "a product of many prime factors." For example, the number $1024 = 2^{10}$ is highly composite since it is a power of 2. The number $360 = 2 \cdot 3^2 \cdot 4 \cdot 5$ is also composite, but it requires prime factors other than 2. Prime numbers $(2, 3, 5, 7, 11, 13, 17, \ldots)$ are not composite at all.

A.1 Mixed-Radix Cooley-Tukey FFT

When the desired DFT length N can be expressed as a product of smaller integers, the Cooley-Tukey decomposition provides what is called a *mixed radix* Cooley-Tukey FFT algorithm.[2]

Two basic varieties of Cooley-Tukey FFT are *decimation in time* (DIT) and its Fourier dual, *decimation in frequency* (DIF). The next section illustrates decimation in time.

A.1.1 Decimation in Time

The DFT is defined by

$$X(k) = \sum_{n=0}^{N-1} x(n) W_N^{kn}, \quad k = 0, 1, 2, \ldots, N-1,$$

where $x(n)$ is the input signal amplitude at time n, and

$$W_N \stackrel{\Delta}{=} e^{-j\frac{2\pi}{N}}. \quad \text{(primitive } N\text{th root of unity)}$$

Note that $W_N^N = 1$.

When N is even, the DFT summation can be split into sums over the odd and even indexes of the input signal:

$$\begin{aligned}
X(\omega_k) &\stackrel{\Delta}{=} \text{DFT}_{N,k}\{x\} \stackrel{\Delta}{=} \sum_{n=0}^{N-1} x(n) e^{-j\omega_k nT}, \quad \omega_k \stackrel{\Delta}{=} \frac{2\pi k}{NT} \\
&= \sum_{\substack{n=0 \\ n \text{ even}}}^{N-2} x(n) e^{-j\omega_k nT} + \sum_{\substack{n=0 \\ n \text{ odd}}}^{N-1} x(n) e^{-j\omega_k nT} \\
&= \sum_{n=0}^{\frac{N}{2}-1} x(2n) e^{-j2\pi \frac{k}{N/2} n} + e^{-j2\pi \frac{k}{N}} \sum_{n=0}^{\frac{N}{2}-1} x(2n+1) e^{-j2\pi \frac{k}{N/2} n}, \\
&= \sum_{n=0}^{\frac{N}{2}-1} x_e(n) W_{N/2}^{kn} + W_N^k \sum_{n=0}^{\frac{N}{2}-1} x_o(n) W_{N/2}^{kn} \\
&\stackrel{\Delta}{=} \text{DFT}_{\frac{N}{2},k}\{\text{Downsample}_2(x)\} \\
&\quad + W_N^k \cdot \text{DFT}_{\frac{N}{2},k}\{\text{Downsample}_2[\text{Shift}_1(x)]\}, \qquad \text{(A.1)}
\end{aligned}$$

where $x_e(n) \stackrel{\Delta}{=} x(2n)$ and $x_o(n) \stackrel{\Delta}{=} x(2n+1)$ denote the even- and odd-indexed samples from x. Thus, the length N DFT is computable using two

[2]See `http://en.wikipedia.org/wiki/Cooley-Tukey_FFT_algorithm`.

length $N/2$ DFTs. The complex factors $W_N^k = e^{-j\omega_k} = \exp(-j2\pi k/N)$ are called *twiddle factors*. The splitting into sums over even and odd time indexes is called *decimation in time*. (For *decimation in frequency*, the inverse DFT of the spectrum $X(\omega_k)$ is split into sums over even and odd *bin numbers* k.)

A.1.2 Radix 2 FFT

When N is a power of 2, say $N = 2^K$ where $K > 1$ is an integer, then the above DIT decomposition can be performed $K - 1$ times, until each DFT is length 2. A length 2 DFT requires no multiplies. The overall result is called a *radix 2 FFT*. A different radix 2 FFT is derived by performing decimation in frequency.

A *split radix* FFT is theoretically more efficient than a pure radix 2 algorithm [70, 29] because it minimizes real arithmetic operations. The term "split radix" refers to a DIT decomposition that combines portions of one radix 2 and two radix 4 FFTs [20].[3] On modern general-purpose processors, however, computation time is often not minimized by minimizing the arithmetic operation count (see §A.7 below).

Radix 2 FFT Complexity is N Log N

Putting together the length N DFT from the $N/2$ length-2 DFTs in a radix-2 FFT, the only multiplies needed are those used to combine two small DFTs to make a DFT twice as long, as in Eq. (A.1). Since there are approximately N (complex) multiplies needed for each stage of the DIT decomposition, and only $\lg N$ stages of DIT (where $\lg N$ denotes the log-base-2 of N), we see that the total number of multiplies for a length N DFT is reduced from $\mathcal{O}(N^2)$ to $\mathcal{O}(N \lg N)$, where $\mathcal{O}(x)$ means "on the order of x". More precisely, a complexity of $\mathcal{O}(N \lg N)$ means that given any implementation of a length-N radix-2 FFT, there exist a constant C and integer M such that the computational complexity $\mathcal{C}(N)$ satisfies

$$\mathcal{C}(N) \leq CN \lg N$$

for all $N > M$. In summary, the complexity of the radix-2 FFT is said to be "N log N", or $\mathcal{O}(N \lg N)$.

[3] `http://en.wikipedia.org/wiki/Split_radix`

A.2 Prime Factor Algorithm (PFA)

By the prime factorization theorem, every integer N can be uniquely factored into a product of prime numbers p_i raised to an integer power $m_i \geq 1$:

$$N = \prod_{i=1}^{n_p} p_i^{m_i}$$

As discussed above, a mixed-radix Cooley Tukey FFT can be used to implement a length N DFT using DFTs of length p_i. However, for factors of N that are mutually prime (such as $p_i^{m_i}$ and $p_j^{m_j}$ for $i \neq j$), a more efficient *prime factor algorithm* (PFA), also called the *Good-Thomas FFT algorithm*, can be used [24, 77, 33, 41, 8, 80].[4] The Chinese Remainder Theorem[5] is used to *re-index* either the input or output samples for the PFA.[6] Since the PFA is only applicable to mutually prime factors of N, it is ideally combined with a mixed-radix Cooley-Tukey FFT, which works for any integer factors.

It is interesting to note that the PFA actually predates the Cooley-Tukey FFT paper of 1965 [14], with Good's 1958 work on the PFA being cited in that paper [80].

The PFA and *Winograd transform* [41] are closely related, with the PFA being somewhat faster [7].

A.3 Rader's FFT Algorithm for Prime Lengths

Rader's FFT algorithm can be used to compute DFTs of length N in $\mathcal{O}(N \lg N)$ operations when N is a prime number. For an introduction, see the Wikipedia page for Rader's FFT Algorithm:

`http://en.wikipedia.org/wiki/Rader's_FFT_algorithm`

A.4 Bluestein's FFT Algorithm

Like Rader's FFT, *Bluestein's FFT algorithm* (also known as the *chirp z-transform algorithm*), can be used to compute prime-length DFTs in $\mathcal{O}(N \lg N)$ operations [22, pp. 213–215].[7] However, unlike Rader's FFT,

---

[4] See `http://en.wikipedia.org/wiki/Prime-factor_FFT_algorithm` for an introduction.

[5] `http://www.cut-the-knot.org/blue/chinese.shtml`

[6] `http://en.wikipedia.org/wiki/Prime-factor_FFT_algorithm#Re-indexing`

[7] See `http://en.wikipedia.org/wiki/Bluestein's_FFT_algorithm` for another derivation and description of Bluestein's FFT algorithm.

Bluestein's algorithm is not restricted to prime lengths, and it can compute other kinds of transforms, as discussed further below.

Beginning with the DFT

$$X(k) = \sum_{n=0}^{N-1} x(n) W^{-kn}, \quad k = 0, 1, 2, \ldots, N-1,$$

where $W \triangleq \exp(j2\pi/N)$ denotes a primitive Nth root of unity,[8] we multiply and divide by $W^{(n^2+k^2)/2}$ to obtain

$$\begin{aligned} X(k) &= \sum_{n=0}^{N-1} x(n) W^{-kn} W^{\frac{1}{2}(n^2+k^2)} W^{-\frac{1}{2}(n^2+k^2)} \\ &= W^{-\frac{1}{2}k^2} \sum_{n=0}^{N-1} \left[x(n) W^{-\frac{1}{2}n^2} \right] W^{\frac{1}{2}(k-n)^2} \\ &= W^{-\frac{1}{2}k^2} (x_q * w_q)_k, \end{aligned}$$

where '$*$' denotes convolution (§7.2.4 on page 121), and the sequences x_q and w_q are defined by

$$\begin{aligned} x_q(n) &\triangleq x(n) W^{-\frac{1}{2}n^2}, \quad n = 0, 1, 2, \ldots, N-1 \\ w_q(n) &\triangleq W^{\frac{1}{2}n^2}, \quad n = -N+1, -N+2, \ldots, -1, 0, 1, 2, \ldots, N-1, \end{aligned}$$

where the ranges of n given are those actually required by the convolution sum above. Beyond these required minimum ranges for n, the sequences may be extended by zeros. As a result, we may implement this convolution (which is cyclic for even N and "negacyclic" for odd N) using zero-padding and a larger cyclic convolution, as mentioned in §7.2.4 on page 121. In particular, the larger cyclic convolution size $N_2 \geq 2N-1$ may be chosen a power of 2, which need not be larger than $4N-3$. Within this larger cyclic convolution, the negative-n indexes map to $N_2 - N + 1 : N_2 - 1$ in the usual way.

Note that the sequence $x_q(n)$ above consists of the original data sequence $x(n)$ multiplied by a signal $\exp(j\pi n^2/N)$ which can be interpreted as a sampled complex sinusoid with instantaneous normalized radian frequency $2\pi n/N$, *i.e.*, an instantaneous frequency that increases linearly with time. Such signals are called *chirp signals*. For this reason, Bluestein's algorithm is also called the chirp z-transform algorithm [57].

In summary, Bluestein's FFT algorithm provides complexity $N \lg N$ for any positive integer DFT-length N whatsoever, even when N is prime.

[8] Note that W is the complex-conjugate of W_N used in §A.1.1 above.

Other adaptations of the Bluestein FFT algorithm can be used to compute a contiguous subset of DFT frequency samples (any uniformly spaced set of samples along the unit circle), with $N \lg N$ complexity. It can similarly compute samples of the z transform along a sampled spiral of the form z^k, where z is any complex number, and $k = k_0, k_0+1, \ldots, k_0+N-1$, again with complexity $N \lg N$ [22].

A.5 Fast Transforms in Audio DSP

Since most audio signal processing applications benefit from *zero padding* (see §8.1 on page 166), in which case we can always choose the FFT length to be a power of 2, there is almost never a need in practice for more "exotic" FFT algorithms than the basic "pruned" power-of-2 algorithms. (Here "pruned" means elimination of all unnecessary operations, such as when the input signal is real [71, 19].)

An exception is when processing exactly periodic signals where the period is known to be an exact integer number of samples in length.[9] In such a case, the DFT of one period of the waveform can be interpreted as a *Fourier series* (§B.3) of the periodic waveform, and unlike virtually all other practical spectrum analysis scenarios, spectral interpolation is not needed (or wanted). In the exactly periodic case, the spectrum is truly zero between adjacent harmonic frequencies, and the DFT of one period provides spectral samples only at the harmonic frequencies.

Adaptive FFT software (see §A.7 below) will choose the fastest algorithm available for any desired DFT length. Due to modern processor architectures, execution time is not normally minimized by minimizing arithmetic complexity [21].

A.6 Related Transforms

There are alternatives to the Cooley-Tukey FFT which can serve the same or related purposes and which can have advantages in certain situations [8]. Examples include the fast discrete cosine transform (DCT) [4], discrete Hartley transform [19], and number theoretic transform [2].

The DCT, used extensively in image coding, is described in §A.6.1 below. The Hartley transform, optimized for processing real signals, does not appear to have any advantages over a "pruned real-only FFT" [71].

[9] Obtaining an exact integer number of samples per period can be arranged using *pitch detection* and *resampling* of the periodic signal. A time-varying pitch requires time-varying resampling [69]—see Appendix D. However, when a signal is resampled for this purpose, one can generally choose a power of 2 for the number of samples per period.

The number theoretic transform has special applicability for large-scale, high-precision calculations (§A.6.2 below).

A.6.1 The Discrete Cosine Transform (DCT)

In image coding (such as MPEG and JPEG), and many audio coding algorithms (MPEG), the *discrete cosine transform* (DCT) is used because of its nearly optimal asymptotic theoretical *coding gain*.[10] For 1D signals, one of several DCT definitions (the one called DCT-II)[11] is given by

$$\begin{aligned}\text{DCT}_k(x) &= 2\sum_{n=0}^{N-1} x(n)\cos\left[\frac{\pi k}{2N}(2n+1)\right], \quad k=0,1,2,\ldots,N-1\\ &= 2\sum_{n=0}^{N-1} x(n)\cos\left[\omega_k\left(n+\frac{1}{2}\right)\right] \qquad \text{(A.2)}\end{aligned}$$

where

$$\omega_k \triangleq \frac{2\pi}{2N}k.$$

Note that ω_k is the DFT frequency for a length $2N$ DFT (as opposed to N).

For real signals, the real part of the DFT is a kind of DCT:

$$\begin{aligned}\text{re}\left\{\text{DFT}_k(x)\right\} &\triangleq \text{re}\left\{\sum_{n=0}^{N-1} x(n)e^{-j\omega_k n}\right\}\\ &= \sum_{n=0}^{N-1} x(n)\text{re}\left\{e^{-j\omega_k n}\right\}\\ &= \sum_{n=0}^{N-1} x(n)\cos(\omega_k n)\end{aligned}$$

Thus, the real part of a double-length FFT is the same as the DCT except for the half-sample phase shift in the sinusoidal basis functions $s_k(n) = e^{j\omega_k n}$ (and a scaling by 2 which is unimportant).

---

[10]This result is well known in the field of image processing. The DCT performs almost as well as the optimal Karhunen-Loève Transform (KLT) when analyzing certain Gaussian stochastic processes as the transform size goes to infinity. (In the KLT, the basis functions are taken to be the eigenvectors of the autocorrelation matrix of the input signal block. As a result, the transform coefficients are *decorrelated* in the KLT, leading to maximum energy concentration and optimal coding gain.) However, the DFT provides a similar degree of optimality for large block sizes N. For practical spectral analysis and processing of audio signals, there is typically no reason to prefer the DCT over the DFT.

[11]For a discussion of eight or so DCT variations, see the Wikipedia page: `http://en.wikipedia.org/wiki/Discrete_cosine_transform`

In practice, the DCT is normally implemented using the same basic efficiency techniques as in FFT algorithms. In Matlab and Octave (Octave-Forge), the functions `dct` and `dct2` are available for the 1D and 2D cases, respectively.

Exercise: Using Euler's identity, expand the cosine in the DCT defined by Eq. (A.2) above into a sum of complex sinusoids, and show that the DCT can be rewritten as the sum of two phase-modulated DFTs:

$$\text{DCT}_{N,k} = e^{j\omega_k \frac{1}{2}}\overline{X_{2N}(\omega_k)} + e^{-j\omega_k \frac{1}{2}} X_{2N}(\omega_k)$$

where $X_{2N}(\omega_k) = \text{DFT}_{2N,k}(x)$ denotes the length $2N$ DFT of x.

A.6.2 Number Theoretic Transform

The *number theoretic transform* is based on generalizing the Nth primitive root of unity (see §3.12 on page 28) to a "quotient ring" instead of the usual field of complex numbers. Let W_N denote a primitive Nth root of unity. We have been using $W_N = \exp(-j2\pi/N)$ in the field of complex numbers, and it of course satisfies $W_N^N = 1$, making it a root of unity; it also has the property that W_N^k visits all of the "DFT frequency points" on the unit circle in the z plane, as k goes from 0 to $N-1$.

In a number theory transform, W_N is an *integer* which satisfies

$$W_N^N = 1 \,(\text{mod}\, p)$$

where p is a prime integer. From number theory, for each prime number p there exists at least one primitive root r such that r^n (modulo p) visits all of the numbers 1 through $p-1$ *in some order*, as n goes from 1 to $p-1$. Since $m^{p-1} = 1\,(\text{mod}\, p)$ for all integers m (another result from number theory), r is also an Nth root of unity, where $N = p-1$ is the transform size. (More generally, N can be any integer divisor L of $p-1$, in which case we use $W_N = r^L$ as the generator of the numbers participating in the transform.)

When the number of elements in the transform is composite, a "fast number theoretic transform" may be constructed in the same manner as a fast Fourier transform is constructed from the DFT, or as the prime factor algorithm (or Winograd transform) is constructed for products of small mutually prime factors [41].

Unlike the DFT, the number theoretic transform does not transform to a meaningful "frequency domain". However, it has analogous theorems, such as the convolution theorem, enabling it to be used for fast convolutions and correlations like the various FFT algorithms.

An interesting feature of the number theory transform is that all computations are *exact* (integer multiplication and addition modulo a prime integer). There is no round-off error. This feature has been used to do fast convolutions to multiply extremely large numbers, such as are needed when computing π to millions of digits of precision.

A.7 FFT Software

For those of us in signal processing research, the built-in `fft` function in Matlab (or Octave) is what we use almost all the time. It is adaptive in that it will choose the best algorithm available for the desired transform size.

For C or C++ applications, there are several highly optimized FFT variants in the `FFTW`[12] package ("Fastest Fourier Transform in the West") [21]. `FFTW` is free for non-commercial or free-software applications under the terms of the GNU General Public License.

For embedded DSP applications (software running on special purpose DSP chips), consult your vendor's software libraries and support website for FFT algorithms written in optimized assembly language for your DSP hardware platform. Nearly all DSP chip vendors supply free FFT software (and other signal processing utilities) as a way of promoting their hardware.

[12] `http://www.fftw.org/`

Appendix B

Fourier Transforms for Continuous/Discrete Time/Frequency

The Fourier transform can be defined for signals which are

- discrete or continuous in time, and
- finite or infinite in duration.

This results in four cases. Quite naturally, the frequency domain has the same four cases,

- discrete or continuous in frequency, and
- finite or infinite in bandwidth.

When time is discrete, the frequency axis is finite, and vice versa.

This book has been concerned almost exclusively with the discrete-time, discrete-frequency case (the DFT), and in that case, both the time and frequency axes are finite in length. In the following sections, we briefly summarize the other three cases. Table B.1 summarizes all four Fourier-transform cases corresponding to discrete or continuous time and/or frequency.

Time Duration		
Finite	Infinite	
Discrete FT (DFT)	Discrete Time FT (DTFT)	discr.
$X(k) = \sum_{n=0}^{N-1} x(n)e^{-j\omega_k n}$	$X(\omega) = \sum_{n=-\infty}^{+\infty} x(n)e^{-j\omega n}$	time
$k = 0, 1, \ldots, N-1$	$\omega \in [-\pi, +\pi)$	n
Fourier Series (FS)	Fourier Transform (FT)	cont.
$X(k) = \frac{1}{P}\int_0^P x(t)e^{-j\omega_k t}dt$	$X(\omega) = \int_{-\infty}^{+\infty} x(t)e^{-j\omega t}dt$	time
$k = -\infty, \ldots, +\infty$	$\omega \in (-\infty, +\infty)$	t
discrete freq. k	continuous freq. ω	

Table B.1: Four cases of sampled/continuous time and frequency.

B.1 Discrete Time Fourier Transform (DTFT)

The *Discrete Time Fourier Transform* (DTFT) can be viewed as the limiting form of the DFT when its length N is allowed to approach infinity:

$$X(\tilde{\omega}) \triangleq \sum_{n=-\infty}^{\infty} x(n)e^{-j\tilde{\omega}n}$$

where $\tilde{\omega} \triangleq \omega T \in [-\pi, \pi)$ denotes the *continuous* normalized radian frequency variable,[1] and $x(n)$ is the signal amplitude at sample number n.

The inverse DTFT is

$$x(n) = \frac{1}{2\pi}\int_{-\pi}^{\pi} X(\tilde{\omega})e^{j\tilde{\omega}n}d\tilde{\omega}$$

which can be derived in a manner analogous to the derivation of the inverse DFT (see Chapter 6).

Instead of operating on sampled signals of length N (like the DFT), the DTFT operates on sampled signals $x(n)$ defined over all integers $n \in \mathbf{Z}$. As a result, the DTFT frequencies form a *continuum*. That is, the DTFT is a function of *continuous* frequency $\tilde{\omega} \in [-\pi, \pi)$, while the DFT is a function of discrete frequency ω_k, $k \in [0, N-1]$. The DFT frequencies $\omega_k = 2\pi k/N$, $k = 0, 1, 2, \ldots, N-1$, are given by the angles of N points

[1] We define the DTFT using *normalized radian frequency* $\tilde{\omega} \triangleq \omega T \in [-\pi, \pi)$.

uniformly distributed along the unit circle in the complex plane (see Fig. 6.1 on page 101). Thus, as $N \to \infty$, a continuous frequency axis must result in the limit along the unit circle in the z plane. The axis is still finite in length, however, because the time domain remains sampled.

B.2 Fourier Transform (FT) and Inverse

The *Fourier transform* of a signal $x(t) \in \mathbf{C}$, $t \in (-\infty, \infty)$, is defined as

$$X(\omega) \triangleq \int_{-\infty}^{\infty} x(t)e^{-j\omega t}dt, \tag{B.1}$$

and its inverse is given by

$$x(t) = \frac{1}{2\pi}\int_{-\infty}^{\infty} X(\omega)e^{j\omega t}d\omega. \tag{B.2}$$

B.2.1 Existence of the Fourier Transform

Conditions for the *existence* of the Fourier transform are complicated to state in general [10], but it is *sufficient* for $x(t)$ to be *absolutely integrable*, *i.e.*,

$$\| x \|_1 \triangleq \int_{-\infty}^{\infty} |x(t)|\, dt < \infty.$$

This requirement can be stated as $x \in L^1$, meaning that x belongs to the set of all signals having a finite L^1 norm ($\| x \|_1 < \infty$). It is similarly sufficient for $x(t)$ to be *square integrable*, *i.e.*,

$$\| x \|_2^2 \triangleq \int_{-\infty}^{\infty} |x(t)|^2\, dt < \infty,$$

or, $x \in L^2$. More generally, it suffices to show $x \in L^p$ for $1 \leq p \leq 2$ [10, p. 47].

There is never a question of existence, of course, for Fourier transforms of real-world signals encountered in practice. However, *idealized* signals, such as sinusoids that go on forever in time, do pose normalization difficulties. In practical engineering analysis, these difficulties are resolved using Dirac's "generalized functions" such as the *impulse* (also called the *delta function*) [36].

B.2.2 The Continuous-Time Impulse

An *impulse* in continuous time must have *"zero width"* and *unit area* under it. One definition is

$$\delta(t) \isdef \lim_{\Delta\to 0} \left\{ \begin{array}{ll} \frac{1}{\Delta}, & 0 \le t \le \Delta \\ 0, & \mbox{otherwise}. \end{array} \right. \tag{B.3}$$

An impulse can be similarly defined as the limit of *any* integrable pulse shape which maintains unit area and approaches zero width at time 0. As a result, the impulse under every definition has the so-called *sifting property* under integration,

$$\int_{-\infty}^{\infty} f(t)\,\delta(t)\,dt = f(0), \tag{B.4}$$

provided $f(t)$ is continuous at $t = 0$. This is often taken as the *defining property* of an impulse, allowing it to be defined in terms of non-vanishing function limits such as

$$\delta(t) \isdef \lim_{\Omega\to\infty} \frac{\sin(\Omega t)}{\pi t}.$$

(Note, incidentally, that $\sin(\Omega t)/\pi t$ is in L^2 but not L^1.)

An impulse is not a function in the usual sense, so it is called instead a *distribution* or *generalized function* [10, 36]. (It is still commonly called a "delta function", however, despite the misnomer.)

B.3 Fourier Series (FS) and Relation to DFT

In continuous time, a *periodic* signal $x(t)$, with period P seconds,[2] may be expanded into a *Fourier series* with coefficients given by

$$X(\omega_k) \isdef \frac{1}{P}\int_0^P x(t) e^{-j\omega_k t} dt, \quad k = 0, \pm 1, \pm 2, \ldots \tag{B.5}$$

where $\omega_k \isdef 2\pi k/P$ is the kth harmonic frequency (rad/sec). The generally complex value $X(\omega_k)$ is called the kth *Fourier series coefficient*. The normalization by $1/P$ is optional, but often included to make the Fourier series coefficients independent of the fundamental frequency $1/P$, and thereby depend only on the *shape* of one period of the time waveform.

[2] A signal $x(t)$ is said to be *periodic* with period P if $x(t + P) = x(t)$ for all $t \in \mathbf{R}$.

B.3.1 Relation of the DFT to Fourier Series

We now show that the DFT of a sampled signal $x(n)$ (of length N), is proportional to the *Fourier series coefficients* of the continuous periodic signal obtained by repeating and interpolating x. More precisely, the DFT of the N samples comprising one period equals N times the Fourier series coefficients. To avoid aliasing upon sampling, the continuous-time signal must be bandlimited to less than half the sampling rate (see Appendix D); this implies that at most N complex harmonic components can be nonzero in the original continuous-time signal.

If $x(t)$ is bandlimited to $\omega T \in (-\pi, \pi)$, it can be sampled at intervals of T seconds without aliasing (see §D.2 on page 223). One way to sample a signal inside an integral expression such as Eq. (B.5) is to multiply it by a continuous-time *impulse train*

$$\Psi_T(t) \stackrel{\Delta}{=} T \sum_{n=-\infty}^{\infty} \delta(t - nT) \tag{B.6}$$

where $\delta(t)$ is the continuous-time impulse signal defined in Eq. (B.3).

We wish to find the continuous-time Fourier series of the *sampled* periodic signal $x(nT)$. Thus, we replace $x(t)$ in Eq. (B.5) by

$$x_s(t) \stackrel{\Delta}{=} x(t) \cdot \Psi_T(t).$$

By the sifting property of delta functions (Eq. (B.4)), the Fourier series of x_s is[3]

$$X_s(\omega_k) = \frac{1}{P}\int_0^P x_s(t)e^{-j\omega_k t}dt \quad = \quad \frac{1}{P}\sum_{n=0}^{\lceil P/T\rceil -1} x(nT)e^{-j\omega_k nT}T.$$

If the sampling interval T is chosen so that it divides the signal period P, then the number of samples under the integral is an integer $N \stackrel{\Delta}{=} P/T$, and we obtain

$$\begin{aligned} X_s(\omega_k) &= \frac{T}{P}\sum_{n=0}^{N-1} x(nT)e^{-j\omega_k nT} \\ &\stackrel{\Delta}{=} \frac{1}{N}\text{DFT}_{N,k}(x_p), \quad k = 0, \pm 1, \pm 2, \dots \end{aligned}$$

where $x_p \stackrel{\Delta}{=} [x(0), x(T), \dots, x((N-1)T)]$. Thus, $X_s(\omega_k) = X(\omega_k)$ for all k at which the bandlimited periodic signal $x(t)$ has a nonzero harmonic.

[3]To obtain precisely this result, it is necessary to define $\delta(t)$ via a limiting pulse converging to time 0 from the *right* of time 0, as we have done in Eq. (B.3).

When N is odd, $X(\omega_k)$ can be nonzero for $k \in [-(N-1)/2, (N-1)/2]$, while for N even, the maximum nonzero harmonic-number range is $k \in [-N/2+1, N/2-1]$.

In summary,

> *the Fourier series coefficients of a periodic, bandlimited signal x are given by the DFT of one period of the samples of x, divided by N, where N is the DFT length, and N is also the number of samples in each period of x.*

Appendix C

Selected Continuous-Time Fourier Theorems

This appendix presents Fourier theorems which are nice to know, but which do not, strictly speaking, pertain to the DFT. The *differentiation theorem* for Fourier Transforms comes up quite often, and its dual pertains as well to the DTFT (Appendix B). The *scaling theorem* provides an important basic insight into time-frequency duality. Finally, the very fundamental *uncertainty principle* is related to the scaling theorem.

C.1 Differentiation Theorem

Let $x(t)$ denote a function differentiable for all t such that $x(\pm\infty) = 0$ and the Fourier Transforms (FT) of both $x(t)$ and $x'(t)$ exist, where $x'(t)$ denotes the time derivative of $x(t)$. Then we have

$$\boxed{x'(t) \longleftrightarrow j\omega X(\omega)}$$

where $X(\omega)$ denotes the Fourier transform of $x(t)$. In operator notation:

$$\boxed{\text{FT}_\omega(x') = j\omega X(\omega)}$$

Proof: This follows immediately from integration by parts:

$$\begin{aligned}
\text{FT}_\omega(x') &\triangleq \int_{-\infty}^{\infty} x'(t)e^{-j\omega t}dt \\
&= x(t)e^{-j\omega t}\Big|_{-\infty}^{\infty} - \int_{-\infty}^{\infty} x(t)(-j\omega)e^{-j\omega t}dt \\
&= j\omega X(\omega)
\end{aligned}$$

since $x(\pm\infty) = 0$.

The differentiation theorem is implicitly used in §E.6 on page 237 to show that audio signals are perceptually equivalent to bandlimited signals which are infinitely differentiable for all time.

C.2 Scaling Theorem

The *scaling theorem* (or *similarity theorem*) provides that if you horizontally "stretch" a signal by the factor α in the time domain, you "squeeze" its Fourier transform by the same factor in the frequency domain. This is an important general Fourier duality relationship.

Theorem: For all continuous-time functions $x(t)$ possessing a Fourier transform,

$$\boxed{\text{SCALE}_\alpha(x) \longleftrightarrow |\alpha|\,\text{SCALE}_{(1/\alpha)}(X)}$$

where

$$\text{SCALE}_{\alpha,t}(x) \triangleq x\left(\frac{t}{\alpha}\right)$$

and α is any nonzero real number (the abscissa stretch factor).

Proof: Taking the Fourier transform of the stretched signals gives

$$\begin{aligned}
\text{FT}_\omega(\text{SCALE}_\alpha(x)) &\triangleq \int_{-\infty}^{\infty} x\left(\frac{t}{\alpha}\right) e^{-j\omega t} dt \qquad (\text{let } \tau = t/\alpha)\\
&= \int_{-\infty}^{\infty} x(\tau) e^{-j\omega(\alpha\tau)} d(\alpha\tau)\\
&= |\alpha| \int_{-\infty}^{\infty} x(\tau) e^{-j(\alpha\omega)\tau} d\tau\\
&\triangleq |\alpha|\, X(\alpha\omega).
\end{aligned}$$

The absolute value appears above because, when $\alpha < 0$, $d(\alpha\tau) < 0$, which brings out a minus sign in front of the integral from $-\infty$ to ∞.

The scaling theorem is fundamentally restricted to the continuous-time, continuous-frequency (Fourier transform) case. The closest we came to the scaling theorem among the DFT theorems was the stretch theorem (§7.4.10 on page 157). We found that "stretching" a *discrete-time* signal by the integer factor α (filling in between samples with zeros) corresponded to the spectrum being *repeated* α times around the unit circle. As a result, the "baseband" copy of the spectrum "shrinks" in width (relative to 2π) by the factor α. Similarly, stretching a signal using *interpolation* (instead of zero-fill) corresponded to the same repeated spectrum with the spectral copies zeroed out. The spectrum of the interpolated signal can therefore be seen as having been stretched by the inverse of the time-domain stretch factor. In summary, the stretch theorem for DFTs can be viewed as the discrete-time, discrete-frequency counterpart of the scaling theorem for Fourier Transforms.

C.3 The Uncertainty Principle

The *uncertainty principle* (for Fourier transform pairs) follows immediately from the scaling theorem. It may be loosely stated as

$$\boxed{\textit{Time Duration} \times \textit{Frequency Bandwidth} \geq c}$$

where c is some constant determined by the precise definitions of "duration" in the time domain and "bandwidth" in the frequency domain.

If duration and bandwidth are defined as the "nonzero interval," then we obtain $c = \infty$, which is not very useful. This conclusion follows immediately from the definition of the Fourier transform and its inverse in §B.2.

C.3.1 Duration and Bandwidth as Second Moments

More interesting definitions of duration and bandwidth are obtained for nonzero signals using the normalized *second moments* of the squared magnitude:

$$\begin{aligned} \Delta t &\stackrel{\Delta}{=} \frac{1}{\|x\|_2}\sqrt{\int_{-\infty}^{\infty} t^2\,|x(t)|^2\,dt} \stackrel{\Delta}{=} \frac{\|tx\|_2}{\|x\|_2} \\ \Delta\omega &\stackrel{\Delta}{=} \frac{1}{\|X\|_2}\sqrt{\int_{-\infty}^{\infty} \omega^2\,|X(\omega)|^2\,\frac{d\omega}{2\pi}} \stackrel{\Delta}{=} \frac{\|\omega X\|_2}{\|X\|_2}, \end{aligned} \tag{C.1}$$

where

$$\begin{aligned} \|x\|_2^2 &\stackrel{\Delta}{=} \int_{-\infty}^{\infty} |x(t)|^2\,dt \\ \|X\|_2^2 &\stackrel{\Delta}{=} \int_{-\infty}^{\infty} |X(\omega)|^2\,\frac{d\omega}{2\pi}. \end{aligned}$$

By the DTFT power theorem, which is proved in a manner analogous to the DFT case in §7.4.8 on page 156, we have $\|x\|_2 = \|X\|_2$. Note that writing "$\| tx \|_2$" and "$\| \omega X \|_2$" is an abuse of notation, but a convenient one. These duration/bandwidth definitions are routinely used in physics, *e.g.*, in connection with the *Heisenberg uncertainty principle*.[1] Under these definitions, we have the following theorem [50, p. 273–274]:

Theorem: If $x(t) \neq 0$ and $\sqrt{|t|}\,x(t) \to 0$ as $|t| \to \infty$, then

$$\boxed{\Delta t \cdot \Delta\omega \geq \frac{1}{2}} \tag{C.2}$$

with equality if and only if

$$x(t) = Ae^{-\alpha t^2}, \quad \alpha > 0, \quad A \neq 0.$$

That is, only the *Gaussian function* (also known as the "bell curve" or "normal curve") achieves the lower bound on the time-bandwidth product.

Proof: Without loss of generality, we may take consider $x(t)$ to be real and normalized to have unit L^2 norm ($\| x \|_2 = 1$). From the Schwarz inequality (see §5.9.3 on page 82 for the discrete-time case),

$$\left| \int_{-\infty}^{\infty} tx(t) \left[\frac{d}{dt} x(t) \right] dt \right|^2 \leq \int_{-\infty}^{\infty} t^2 x^2(t) dt \int_{-\infty}^{\infty} \left| \frac{d}{dt} x(t) \right|^2 dt. \tag{C.3}$$

The left-hand side can be evaluated using integration by parts:

$$\int_{-\infty}^{\infty} tx \frac{dx}{dt} dt = t \frac{x^2(t)}{2} \bigg|_{-\infty}^{\infty} - \frac{1}{2} \int_{-\infty}^{\infty} x^2(t) dt \triangleq -\frac{1}{2} \| x \|_2^2 = -\frac{1}{2}$$

where we used the assumption that $\sqrt{|t|}\,x(t) \to 0$ as $|t| \to \infty$.

The second term on the right-hand side of Eq. (C.3) can be evaluated using the power theorem (§7.4.8 on page 156 proves the discrete-time case) and differentiation theorem (§C.1 above):

$$\int_{-\infty}^{\infty} \left| \frac{dx(t)}{dt} \right|^2 dt = \int_{-\infty}^{\infty} |j\omega X(\omega)|^2 \frac{d\omega}{2\pi} = \int_{-\infty}^{\infty} \omega^2 |X(\omega)|^2 \frac{d\omega}{2\pi}$$

[1] The Heisenberg uncertainty principle in quantum physics applies to any dual properties of a particle. For example, the position and velocity of an electron are oft-cited as such duals. An electron is described, in quantum mechanics, by a probability wave packet. Therefore, the *position* of an electron in space can be defined as the midpoint of the amplitude envelope of its wave function; its *velocity*, on the other hand, is determined by the *frequency* of the wave packet. To accurately measure the frequency, the packet must be very long in space, to provide many cycles of oscillation under the envelope. But this means the location in space is relatively uncertain. In more precise mathematical terms, the probability wave function for velocity is proportional to the spatial Fourier transform of the probability wave for position. *I.e.*, they are exact Fourier duals. The Heisenberg Uncertainty Principle is therefore a Fourier property of fundamental particles described by waves [17]. This of course includes all matter and energy in the Universe.

Substituting these evaluations into Eq. (C.3) gives

$$\left|-\frac{1}{2}\right|^2 \leq \| tx \|_2^2 \| \omega X \|_2^2 .$$

Taking the square root of both sides gives the uncertainty relation sought.

If equality holds in the uncertainty relation Eq. (C.2), then Eq. (C.3) implies

$$\frac{d}{dt}x(t) = ctx(t)$$

for some constant c, which implies $x(t) = Ae^{\frac{c}{2}t^2}$ for some constants A and c. Since $x(\pm\infty) = 0$ by hypothesis, we have $c < 0$ while $A \neq 0$ remains arbitrary.

C.3.2 Time-Limited Signals

If $x(t) = 0$ for $|t| \geq \Delta t/2$, then

$$\Delta t \cdot \Delta\omega \geq \pi$$

where $\Delta\omega$ is as defined above in Eq. (C.1).

Proof: See [50, pp. 274–5].

C.3.3 Time-Bandwidth Products are Unbounded Above

We have considered two lower bounds for the time-bandwidth product based on two different definitions of duration in time. In the opposite direction, there is *no upper bound* on time-bandwidth product. To see this, imagine filtering an arbitrary signal with an *allpass filter*.[2] The allpass filter cannot affect bandwidth $\Delta\omega$, but the duration Δt can be arbitrarily extended by successive applications of the allpass filter.

[2] An allpass filter has unity gain and arbitrary delay at each frequency.

Appendix D

Sampling Theory

In this appendix, sampling theory is derived as an application of the DTFT and the Fourier theorems developed in Appendix C. First, we must derive a formula for *aliasing* due to uniformly sampling a continuous-time signal. Next, the *sampling theorem* is proved. The sampling theorem provides that a properly bandlimited continuous-time signal can be sampled and reconstructed from its samples without error, in principle.

An early derivation of the sampling theorem is often cited as a 1928 paper by Harold Nyquist, and Claude Shannon is credited with reviving interest in the sampling theorem after World War II when computers became public.[1] As a result, the sampling theorem is often called "Nyquist's sampling theorem," "Shannon's sampling theorem," or the like. Also, the sampling rate has been called the *Nyquist rate* in honor of Nyquist's contributions [46]. In the author's experience, however, modern usage of the term "Nyquist rate" refers instead to *half* the sampling rate. To resolve this clash between historical and current usage, the term *Nyquist limit* will always mean *half* the sampling rate in this book series, and the term "Nyquist rate" will not be used at all.

[1] `http://cnx.org/content/m0050/latest/`

D.1 Introduction to Sampling

Inside computers and modern "digital" synthesizers, (as well as music CDs), sound is *sampled* into a stream of *numbers*. Each *sample* can be thought of as a number which specifies the position[2] of a loudspeaker at a particular instant. When sound is sampled, we call it *digital audio*. The sampling rate used for CDs nowadays is 44,100 samples per second. That means when you play a CD, the speakers in your stereo system are moved to a new position 44,100 times per second, or once every 23 microseconds. Controlling a speaker this fast enables it to generate any sound in the human hearing range because we cannot hear frequencies higher than around 20,000 cycles per second, and a sampling rate more than twice the highest frequency in the sound guarantees that exact reconstruction is possible from the samples.

D.1.1 Reconstruction from Samples—Pictorial Version

Figure D.1 shows how a sound is reconstructed from its samples. Each sample can be considered as specifying the *scaling* and *location* of a *sinc function*. The discrete-time signal being interpolated in the figure is a *digital rectangular pulse*:

$$x = [\ldots, 0, 0, 0, 1, 1, 1, 1, 1, 0, 0, 0, \ldots]$$

The sinc functions are drawn with dashed lines, and they sum to produce the solid curve. An isolated sinc function is shown in Fig. D.2. Note the "Gibb's overshoot" near the corners of the continuous rectangular pulse in Fig. D.1 due to bandlimiting. (A true continuous rectangular pulse has infinite bandwidth.)

[2]More typically, each sample represents the instantaneous *velocity* of the speaker. Here's why: Most microphones are transducers from *acoustic pressure* to *electrical voltage*, and analog-to-digital converters (ADCs) produce numerical samples which are proportional to voltage. Thus, digital samples are normally proportional to *acoustic pressure deviation* (force per unit area on the microphone, with ambient air pressure subtracted out). When digital samples are converted to analog form by digital-to-analog conversion (DAC), each sample is converted to an electrical voltage which then drives a loudspeaker (in audio applications). Typical loudspeakers use a "voice-coil" to convert applied voltage to electromotive force on the speaker which applies pressure on the air via the speaker cone. Since the acoustic impedance of air is a real number, wave pressure is directly proportional wave *velocity*. Since the speaker must move in contact with the air during wave generation, we may conclude that digital signal samples correspond most closely to the *velocity* of the speaker, not its position. The situation is further complicated somewhat by the fact that typical speakers do not themselves have a real driving-point impedance. However, for an "ideal" microphone and speaker, we should get samples proportional to speaker velocity and hence to air pressure. Well below resonance, the real part of the radiation impedance of the pushed air should dominate, as long as the excursion does not exceed the linear interval of cone displacement.

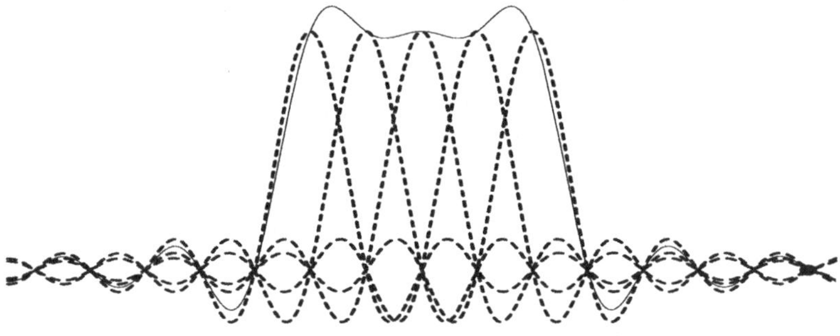

Figure D.1: Summation of weighted sinc functions to create a continuous waveform from discrete-time samples.

Notice that each sinc function passes through zero at every sample instant but the one it is centered on, where it passes through 1.

D.1.2 The Sinc Function

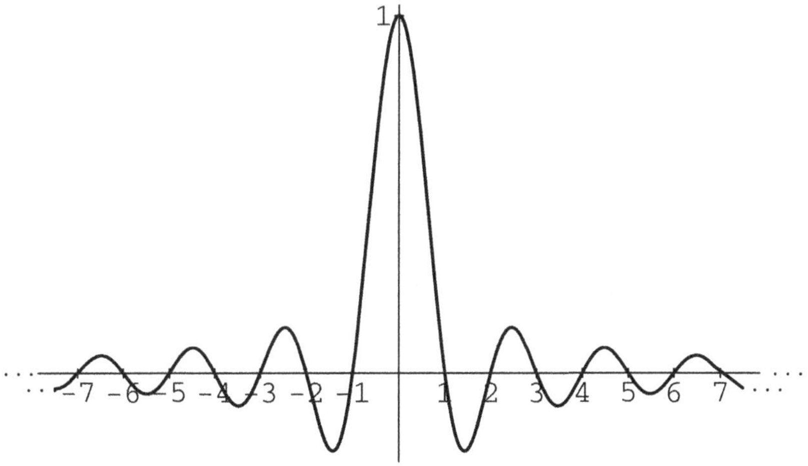

Figure D.2: The sinc function $\text{sinc}(x) \triangleq \sin(\pi x)/(\pi x)$.

The sinc function, or *cardinal sine* function, is the famous "sine x over x" curve, and is illustrated in Fig. D.2. For bandlimited interpolation of discrete-time signals, the ideal *interpolation kernel* is proportional to the sinc function

$$\text{sinc}(f_s t) \triangleq \frac{\sin(\pi f_s t)}{\pi f_s t}.$$

where f_s denotes the sampling rate in samples-per-second (Hz), and t denotes time in seconds. Note that the sinc function has zeros at all the integers except 0, where it is 1. For precise scaling, the desired interpolation kernel is $f_s\text{sinc}(f_s t)$, which has a algebraic area (time integral) that is independent of the sampling rate f_s.

D.1.3 Reconstruction from Samples—The Math

Let $x_d(n) \triangleq x(nT)$ denote the nth sample of the original sound $x(t)$, where t is time in seconds. Thus, n ranges over the integers, and T is the *sampling interval* in seconds. The *sampling rate* in Hertz (Hz) is just the reciprocal of the sampling period, *i.e.*,

$$f_s \triangleq \frac{1}{T}.$$

To avoid losing any information as a result of sampling, we must assume $x(t)$ is *bandlimited* to less than half the sampling rate. This means there can be no energy in $x(t)$ at frequency $f_s/2$ or above. We will prove this mathematically when we prove the *sampling theorem* in §D.3 below.

Let $X(\omega)$ denote the Fourier transform of $x(t)$, *i.e.*,

$$X(\omega) \triangleq \int_{-\infty}^{\infty} x(t)e^{-j\omega t}dt.$$

Then we can say x is *bandlimited* to less than half the sampling rate if and only if $X(\omega) = 0$ for all $|\omega| \geq \pi f_s$. In this case, the sampling theorem gives us that $x(t)$ can be uniquely reconstructed from the samples $x(nT)$ by summing up shifted, scaled, sinc functions:

$$\hat{x}(t) \triangleq \sum_{n=-\infty}^{\infty} x(nT)h_s(t-nT) \equiv x(t)$$

where

$$h_s(t) \triangleq \text{sinc}(f_s t) \triangleq \frac{\sin(\pi f_s t)}{\pi f_s t}.$$

The sinc function is the impulse response of the *ideal lowpass filter*. This means its Fourier transform is a rectangular window in the frequency domain. The particular sinc function used here corresponds to the ideal lowpass filter which cuts off at half the sampling rate. In other words, it has a gain of 1 between frequencies 0 and $f_s/2$, and a gain of zero at all higher frequencies.

The reconstruction of a sound from its samples can thus be interpreted as follows: convert the sample stream into a *weighted impulse train*, and pass that signal through an ideal lowpass filter which cuts off at half the sampling rate. These are the fundamental steps of *digital to analog conversion* (DAC). In practice, neither the impulses nor the lowpass filter are ideal, but they are usually close enough to ideal that one cannot hear any difference. Practical lowpass-filter design is discussed in the context of *bandlimited interpolation*[3] [69].

D.2 Aliasing of Sampled Signals

This section quantifies aliasing in the general case. This result is then used in the proof of the sampling theorem in the next section.

It is well known that when a continuous-time signal contains energy at a frequency higher than half the sampling rate $f_s/2$, sampling at f_s samples per second causes that energy to *alias* to a lower frequency. If we write the original frequency as $f = f_s/2 + \epsilon$, then the new aliased frequency is $f_a = f_s/2 - \epsilon$, for $\epsilon \leq f_s/2$. This phenomenon is also called "folding", since f_a is a "mirror image" of f about $f_s/2$. As we will see, however, this is not a complete description of aliasing, as it only applies to real signals. For general (complex) signals, it is better to regard the aliasing due to sampling as a summation over all spectral "blocks" of width f_s.

Continuous-Time Aliasing Theorem

Let $x(t)$ denote any continuous-time signal having a Fourier Transform (FT)

$$X(j\omega) \stackrel{\Delta}{=} \int_{-\infty}^{\infty} x(t)e^{-j\omega t}dt.$$

Let

$$x_d(n) \stackrel{\Delta}{=} x(nT), \quad n = \dots, -2, -1, 0, 1, 2, \dots,$$

denote the samples of $x(t)$ at uniform intervals of T seconds, and denote its Discrete-Time Fourier Transform (DTFT) by

$$X_d(e^{j\theta}) \stackrel{\Delta}{=} \sum_{n=-\infty}^{\infty} x_d(n)e^{-j\theta n}.$$

---

[3] `http://ccrma.stanford.edu/~jos/resample/`

Then the spectrum X_d of the sampled signal x_d is related to the spectrum X of the original continuous-time signal x by

$$X_d(e^{j\theta}) = \frac{1}{T} \sum_{m=-\infty}^{\infty} X\left[j\left(\frac{\theta}{T} + m\frac{2\pi}{T}\right)\right].$$

The terms in the above sum for $m \neq 0$ are called *aliasing terms.* They are said to *alias* into the *base band* $[-\pi/T, \pi/T]$. Note that the summation of a spectrum with aliasing components involves addition of complex numbers; therefore, aliasing components can be removed only if both their *amplitude and phase* are known.

Proof: Writing $x(t)$ as an inverse FT gives

$$x(t) = \frac{1}{2\pi} \int_{-\infty}^{\infty} X(j\omega) e^{j\omega t} d\omega.$$

Writing $x_d(n)$ as an inverse DTFT gives

$$x_d(n) = \frac{1}{2\pi} \int_{-\pi}^{\pi} X_d(e^{j\theta}) e^{j\theta t} d\theta$$

where $\theta \triangleq 2\pi\omega_d T$ denotes the normalized discrete-time frequency variable.

The inverse FT can be broken up into a sum of finite integrals, each of length $\Omega_s \triangleq 2\pi f_s = 2\pi/T$, as follows:

$$\begin{aligned}
x(t) &= \frac{1}{2\pi} \int_{-\infty}^{\infty} X(j\omega) e^{j\omega t} d\omega \\
&= \frac{1}{2\pi} \sum_{m=-\infty}^{\infty} \int_{(2m-1)\pi/T)}^{(2m+1)\pi/T} X(j\omega) e^{j\omega t} d\omega \qquad (\text{let } \omega \leftarrow m\Omega_s) \\
&= \frac{1}{2\pi} \sum_{m=-\infty}^{\infty} \int_{-\Omega_s/2}^{\Omega_s/2} X\left(j\omega + jm\Omega_s\right) e^{j\omega t} e^{j\Omega_s mt} d\omega \\
&= \frac{1}{2\pi} \int_{-\Omega_s/2}^{\Omega_s/2} e^{j\omega t} \sum_{m=-\infty}^{\infty} X\left(j\omega + jm\Omega_s\right) e^{j\Omega_s mt} d\omega
\end{aligned}$$

Let us now sample this representation for $x(t)$ at $t = nT$ to obtain $x_d(n) \triangleq$

$x(nT)$, and we have

$$\begin{aligned} x_d(n) &= \frac{1}{2\pi}\int_{-\Omega_s/2}^{\Omega_s/2} e^{j\omega nT} \sum_{m=-\infty}^{\infty} X\left(j\omega + jm\Omega_s\right) e^{j\Omega_s mnT} d\omega \\ &= \frac{1}{2\pi}\int_{-\Omega_s/2}^{\Omega_s/2} e^{j\omega nT} \sum_{m=-\infty}^{\infty} X\left(j\omega + jm\Omega_s\right) e^{j(2\pi/T)mnT} d\omega \\ &= \frac{1}{2\pi}\int_{-\Omega_s/2}^{\Omega_s/2} e^{j\omega nT} \sum_{m=-\infty}^{\infty} X\left(j\omega + jm\Omega_s\right) d\omega \end{aligned}$$

since n and m are integers. Normalizing frequency as $\theta' = \omega T$ yields

$$x_d(n) = \frac{1}{2\pi}\int_{-\pi}^{\pi} e^{j\theta' n} \frac{1}{T} \sum_{m=-\infty}^{\infty} X\left[j\left(\frac{\theta'}{T} + m\frac{2\pi}{T}\right)\right] d\theta'.$$

Since this is formally the inverse DTFT of $X_d(e^{j\theta'})$ written in terms of $X(j\theta'/T)$, the result follows.
□

D.3 Sampling Theorem

Let $x(t)$ denote any continuous-time signal having a *continuous* Fourier transform

$$X(j\omega) \triangleq \int_{-\infty}^{\infty} x(t)e^{-j\omega t}dt.$$

Let

$$x_d(n) \triangleq x(nT), \quad n = \ldots, -2, -1, 0, 1, 2, \ldots,$$

denote the samples of $x(t)$ at uniform intervals of T seconds. Then $x(t)$ can be exactly reconstructed from its samples $x_d(n)$ if $X(j\omega) = 0$ for all $|\omega| \geq \pi/T$.[4]

Proof: From the continuous-time aliasing theorem (§D.2), we have that the discrete-time spectrum $X_d(e^{j\theta})$ can be written in terms of the continuous-time spectrum $X(j\omega)$ as

$$X_d(e^{j\omega_d T}) = \frac{1}{T} \sum_{m=-\infty}^{\infty} X[j(\omega_d + m\Omega_s)]$$

[4] Mathematically, $X(j\omega)$ can be allowed to be nonzero over points $|\omega| \geq \pi/T$ provided that the set of all such points have *measure zero* in the sense of Lebesgue integration. However, such distinctions do not arise for practical signals which are always finite in extent and which therefore have continuous Fourier transforms. This is why we specialize the sampling theorem to the case of continuous-spectrum signals.

where $\omega_d \in (-\pi/T, \pi/T)$ is the "digital frequency" variable. If $X(j\omega) = 0$ for all $|\omega| \geq \Omega_s/2$, then the above infinite sum reduces to one term, the $m = 0$ term, and we have

$$X_d(e^{j\omega_d T}) = \frac{1}{T}X(j\omega_d), \quad \omega_d \in \left(-\frac{\pi}{T}, \frac{\pi}{T}\right)$$

At this point, we can see that the spectrum of the sampled signal $x(nT)$ coincides with the nonzero spectrum of the continuous-time signal $x(t)$. In other words, the DTFT of $x(nT)$ is equal to the FT of $x(t)$ between plus and minus half the sampling rate, and the FT is zero outside that range. This makes it clear that spectral information is preserved, so it should now be possible to go from the samples back to the continuous waveform without error, which we now pursue.

To reconstruct $x(t)$ from its samples $x(nT)$, we may simply take the inverse Fourier transform of the zero-extended DTFT, because

$$\begin{aligned} x(t) &= \text{IFT}_t(X) \triangleq \frac{1}{2\pi}\int_{-\infty}^{\infty} X(j\omega)e^{j\omega t}d\omega = \frac{1}{2\pi}\int_{-\Omega_s/2}^{\Omega_s/2} X(j\omega)e^{j\omega t}d\omega \\ &= \frac{1}{2\pi}\int_{-\Omega_s/2}^{\Omega_s/2} X_d(e^{j\theta})e^{j\omega t}d\omega \triangleq \text{IDTFT}_t(X_d). \end{aligned}$$

By expanding $X_d(e^{j\theta})$ as the DTFT of the samples $x(n)$, the formula for reconstructing $x(t)$ as a superposition of weighted sinc functions is obtained (depicted in Fig. D.1):

$$\begin{aligned} x(t) &= \text{IDTFT}_t(X_d) \\ &\triangleq \frac{1}{2\pi}\int_{-\pi}^{\pi} X_d(e^{j\theta})e^{j\omega t}d\omega \\ &= \frac{T}{2\pi}\int_{-\pi/T}^{\pi/T} X_d(e^{j\omega_d T})e^{j\omega_d t}d\omega_d \\ &\triangleq \frac{T}{2\pi}\int_{-\pi/T}^{\pi/T} \left[\sum_{n=-\infty}^{\infty} x(nT)e^{-j\omega_d nT}\right] e^{j\omega_d t}d\omega_d \\ &= \sum_{n=-\infty}^{\infty} x(nT) \underbrace{\frac{T}{2\pi}\int_{-\pi/T}^{\pi/T} e^{j\omega_d(t-nT)}d\omega_d}_{\triangleq h(t-nT)} \\ &\triangleq \sum_{n=-\infty}^{\infty} x(nT)h(t-nT) \\ &\triangleq (x * h)(t) \end{aligned}$$

where we defined

$$\begin{aligned} h(t-nT) &\isdef \frac{T}{2\pi}\int_{-\pi/T}^{\pi/T} e^{j\omega_d(t-nT)}d\omega_d \\ &= \frac{T}{2\pi}\frac{2}{2j(t-nT)}\left[e^{j\pi\frac{t-nT}{T}} - e^{-j\pi\frac{t-nT}{T}}\right] \\ &= \frac{\sin\left[\pi\left(\frac{t}{T}-n\right)\right]}{\pi\left(\frac{t}{T}-n\right)} \\ &\isdef \text{sinc}\left(\frac{t-nT}{T}\right) = \text{sinc}\left(\frac{t}{T}-n\right), \end{aligned}$$

or

$$h(t) = \text{sinc}\left(\frac{t}{T}\right), \quad \text{where sinc}(t) \isdef \frac{\sin(\pi t)}{\pi t}.$$

The "sinc function" is defined with π in its argument so that it has zero crossings on the nonzero integers, and its peak magnitude is 1. Figure D.2 on page 221 illustrates the appearance of the sinc function.

We have shown that when $x(t)$ is bandlimited to less than half the sampling rate, the IFT of the zero-extended DTFT of its samples $x(nT)$ gives back the original continuous-time signal $x(t)$. This completes the proof of the sampling theorem.

□

Conversely, if $x(t)$ can be reconstructed from its samples $x_d(n) \isdef x(nT)$, it must be true that $x(t)$ is bandlimited to $[-f_s/2, f_s/2]$, since a sampled signal only supports frequencies up to $f_s/2$ (see §D.4 below). While a real digital signal $x(n)$ may have energy at half the sampling rate (frequency $f_s/2$), the phase is constrained to be either 0 or π there, which is why this frequency had to be excluded from the sampling theorem.

A one-line summary of the essence of the sampling-theorem proof is

$$\boxed{x(t) = \text{IFT}_t\left\{\text{ZeroPad}_\infty\left\{\text{DTFT}\{x_d\}\right\}\right\}}$$

where $x_d \isdef \text{Sample}_T(x)$.

The sampling theorem is easier to show when applied to sampling-rate conversion in discrete-time, *i.e.*, when simple downsampling of a discrete time signal is being used to reduce the sampling rate by an integer factor. In analogy with the continuous-time aliasing theorem of §D.2, the downsampling theorem (§7.4.11 on page 158) states that downsampling a digital signal by an integer factor L produces a digital signal whose spectrum can be calculated by partitioning the original spectrum into L equal blocks and then summing (aliasing) those blocks. If only one of the blocks is nonzero, then the original signal at the higher sampling rate is exactly recoverable.

D.4 Appendix: Frequencies Representable by a Geometric Sequence

Consider $z_0 \in \mathbf{C}$, with $|z_0| = 1$. Then we can write z_0 in polar form as

$$z_0 \isdef e^{j\theta_0} \isdef e^{j\omega_0 T},$$

where θ_0, ω_0, and T are real numbers.

Forming a geometric sequence based on z_0 yields the sequence

$$x(t_n) \isdef z_0^n = e^{j\theta_0 n} = e^{j\omega_0 t_n}$$

where $t_n \isdef nT$. Thus, successive integer powers of z_0 produce a *sampled complex sinusoid* with unit amplitude, and zero phase. Defining the *sampling interval* as T in seconds provides that ω_0 is the *radian frequency* in radians per second.

A natural question to investigate is what frequencies ω_0 are possible. The angular step of the point z_0^n along the unit circle in the complex plane is $\theta_0 = \omega_0 T$. Since $e^{j(\theta_0 n + 2\pi)} = e^{j\theta_0 n}$, an angular step $\theta_0 > 2\pi$ is indistinguishable from the angular step $\theta_0 - 2\pi$. Therefore, we must restrict the angular step θ_0 to a length 2π range in order to avoid ambiguity.

Recall from §4.3.3 on page 44 that we need support for both positive and negative frequencies since equal magnitudes of each must be summed to produce real sinusoids, as indicated by the identities

$$\begin{aligned}
\cos(\omega_0 t_n) &= \frac{1}{2}e^{j\omega_0 t_n} + \frac{1}{2}e^{-j\omega_0 t_n} \\
\sin(\omega_0 t_n) &= -\frac{j}{2}e^{j\omega_0 t_n} + \frac{j}{2}e^{-j\omega_0 t_n}.
\end{aligned}$$

The length 2π range which is symmetric about zero is

$$\theta_0 \in [-\pi, \pi],$$

which, since $\theta_0 = \omega_0 T = 2\pi f_0 T$, corresponds to the frequency range

$$\begin{aligned}
\omega_0 &\in \left[-\frac{\pi}{T}, \frac{\pi}{T}\right] \\
f_0 &\in \left[-\frac{f_s}{2}, \frac{f_s}{2}\right].
\end{aligned}$$

However, there is a problem with the point at $f_0 = \pm f_s/2$: Both $+f_s/2$ and $-f_s/2$ correspond to the same point $z_0 = -1$ in the z-plane. Technically, this can be viewed as *aliasing* of the frequency $-f_s/2$ on top of $f_s/2$, or vice versa. The practical impact is that it is not possible in general to reconstruct a sinusoid from its samples at this frequency. For an obvious example, consider the sinusoid at half the sampling-rate sampled on its zero-crossings: $\sin(\omega_0 t_n) = \sin(\pi n) \equiv 0$. We cannot be expected to reconstruct a nonzero signal from a sequence of zeros! For the signal $\cos(\omega_0 t_n) = \cos(\pi n) = (-1)^n$, on the other hand, we sample the positive and negative peaks, and everything looks fine. In general, we either do not know the amplitude, or we do not know phase of a sinusoid sampled at exactly twice its frequency, and if we hit the zero crossings, we lose it altogether.

In view of the foregoing, we may define the valid range of "digital frequencies" to be

$$\begin{aligned}\theta_0 &\in [-\pi, \pi) \\ \omega_0 &\in [-\pi/T, \pi/T) \\ f_0 &\in [-f_s/2, f_s/2).\end{aligned}$$

While one might have expected the open interval $(-\pi, \pi)$, we are free to give the point $z_0 = -1$ either the largest positive or largest negative representable frequency. Here, we chose the largest negative frequency since it corresponds to the assignment of numbers in two's complement arithmetic, which is often used to implement phase registers in sinusoidal oscillators. Since there is no corresponding positive-frequency component, samples at $f_s/2$ must be interpreted as samples of clockwise circular motion around the unit circle at two points per revolution. Such signals appear as an alternating sequence of the form $c(-1)^n$, where c can be complex. The amplitude at $-f_s/2$ is then defined as $|c|$, and the phase is $\angle c$.

We have seen that uniformly spaced samples can represent frequencies f_0 only in the range $[-f_s/2, f_s/2)$, where f_s denotes the sampling rate. Frequencies outside this range yield sampled sinusoids indistinguishable from frequencies inside the range.

Suppose we henceforth agree to sample at *higher* than twice the highest frequency in our continuous-time signal. This is normally ensured in practice by lowpass-filtering the input signal to remove all signal energy at $f_s/2$ and above. Such a filter is called an *anti-aliasing filter*, and it is a standard first stage in an *Analog-to-Digital (A/D) Converter (ADC)*. Nowadays, ADCs are normally implemented within a single integrated circuit chip, such as a CODEC (for "coder/decoder") or "multimedia chip".

Appendix E

Taylor Series Expansions

A *Taylor series expansion* of a continuous function $f(x)$ is a *polynomial approximation* of $f(x)$. This appendix derives the Taylor series approximation informally, then introduces the remainder term and a formal statement of Taylor's theorem. Finally, a basic result on the completeness of polynomial approximation is stated.

E.1 Informal Derivation of Taylor Series

We have a function $f(x)$ and we want to approximate it using an nth-order *polynomial*:

$$f(x) = f_0 + f_1 x + f_2 x^2 + \cdots + f_n x^n + R_{n+1}(x)$$

where $R_{n+1}(x)$, the approximation error, is called the *remainder term*. We may assume x and $f(x)$ are *real*, but the following derivation generalizes unchanged to the complex case.

Our problem is to find fixed constants $\{f_i\}_{i=0}^{n}$ so as to obtain the best approximation possible. Let's proceed optimistically as though the approximation will be perfect, and assume $R_{n+1}(x) = 0$ for all x ($R_{n+1}(x) \equiv 0$), given the right values of f_i. Then at $x = 0$ we must have

$$f(0) = f_0$$

That's one constant down and $n - 1$ to go! Now let's look at the first derivative of $f(x)$ with respect to x, again assuming that $R_{n+1}(x) \equiv 0$:

$$f'(x) = 0 + f_1 + 2f_2 x + 3f_2 x^2 + \cdots + n f_n x^{n-1}$$

Evaluating this at $x = 0$ gives

$$f'(0) = f_1.$$

In the same way, we find

$$\begin{aligned} f''(0) &= 2 \cdot f_2 \\ f'''(0) &= 3 \cdot 2 \cdot f_3 \\ &\cdots \\ f^{(n)}(0) &= n! \cdot f_n \end{aligned}$$

where $f^{(n)}(0)$ denotes the nth derivative of $f(x)$ with respect to x, evaluated at $x = 0$. Solving the above relations for the desired constants yields

$$\begin{aligned} f_0 &= f(0) \\ f_1 &= \frac{f'(0)}{1} \\ f_2 &= \frac{f''(0)}{2 \cdot 1} \\ f_3 &= \frac{f'''(0)}{3 \cdot 2 \cdot 1} \\ &\cdots \\ f_n &= \frac{f^{(n)}(0)}{n!}. \end{aligned}$$

Thus, defining $0! \triangleq 1$ (as it always is), we have derived the following polynomial approximation:

$$\boxed{f(x) \approx \sum_{k=0}^{n} \frac{f^{(k)}(0)}{k!} x^k}$$

This is the nth-order Taylor series expansion of $f(x)$ about the point $x = 0$. Its derivation was quite simple. The hard part is showing that the approximation error (remainder term $R_{n+1}(x)$) is small over a wide interval of x values. Another "math job" is to determine the conditions under which the approximation error approaches zero for all x as the order n goes to infinity. The main point to note here is that the Taylor series itself is simple to derive.

E.2 Taylor Series with Remainder

We repeat the derivation of the preceding section, but this time we treat the error term more carefully.

Again we want to approximate $f(x)$ with an nth-order *polynomial*:

$$f(x) = f_0 + f_1 x + f_2 x^2 + \cdots + f_n x^n + R_{n+1}(x)$$

$R_{n+1}(x)$ is the "remainder term" which we will no longer assume is zero.

Our problem is to find $\{f_i\}_{i=0}^n$ so as to minimize $R_{n+1}(x)$ over some interval I containing x. There are many "optimality criteria" we could choose. The one that falls out naturally here is called *Padé approximation*. Padé approximation sets the error value and its first n derivatives to zero at a single chosen point, which we take to be $x = 0$. Since all $n+1$ "degrees of freedom" in the polynomial coefficients f_i are used to set derivatives to zero at one point, the approximation is termed *maximally flat* at that point. In other words, as $x \to 0$, the nth order polynomial approximation approaches $f(x)$ with an error that is proportional to $|x|^{n+1}$.

Padé approximation comes up elsewhere in signal processing. For example, it is the sense in which Butterworth filters are optimal [51]. (Their frequency responses are maximally flat in the center of the pass-band.) Also, Lagrange interpolation filters (which are nonrecursive, while Butterworth filters are recursive), can be shown to maximally flat at dc in the frequency domain [79, 34].

Setting $x = 0$ in the above polynomial approximation produces

$$f(0) = f_0 + R_{n+1}(0) = f_0$$

where we have used the fact that the error is to be exactly zero at $x = 0$ in Padé approximation.

Differentiating the polynomial approximation and setting $x = 0$ gives

$$f'(0) = f_1 + R'_{n+1}(0) = f_1$$

where we have used the fact that we also want the *slope* of the error to be exactly zero at $x = 0$.

In the same way, we find

$$f^{(k)}(0) = k! \cdot f_k + R^{(k)}_{n+1}(0) = k! \cdot f_k$$

for $k = 2, 3, 4, \ldots, n$, and the first n derivatives of the remainder term are all zero. Solving these relations for the desired constants yields the nth-order Taylor series expansion of $f(x)$ about the point $x = 0$

$$f(x) = \sum_{k=0}^{n} \frac{f^{(k)}(0)}{k!} x^k + R_{n+1}(x)$$

as before, but now we better understand the remainder term.

From this derivation, it is clear that the approximation error (remainder term) is smallest in the vicinity of $x = 0$. *All degrees of freedom* in the polynomial coefficients were devoted to minimizing the approximation error and its derivatives at $x = 0$. As you might expect, the approximation error generally worsens as x gets farther away from 0.

To obtain a more *uniform* approximation over some interval I in x, other kinds of error criteria may be employed. Classically, this topic has been called "economization of series," or simply polynomial approximation under different error criteria. In `Matlab` or `Octave`, the function `polyfit(x,y,n)` will find the coefficients of a polynomial $p(x)$ of degree `n` that fits the data `y` over the points `x` in a least-squares sense. That is, it minimizes

$$\| R_{n+1} \|^2 \triangleq \sum_{i=1}^{n_x} |y(i) - p(x(i))|^2$$

where $n_x \triangleq$ `length(x)`.

E.3 Formal Statement of Taylor's Theorem

Let $f(x)$ be continuous on a real interval I containing x_0 (and x), and let $f^{(n)}(x)$ exist at x and $f^{(n+1)}(\xi)$ be continuous for all $\xi \in I$. Then we have the following Taylor series expansion:

$$\begin{aligned} f(x) = f(x_0) &+ \frac{1}{1} f'(x_0)(x - x_0) \\ &+ \frac{1}{1 \cdot 2} f''(x_0)(x - x_0)^2 \\ &+ \frac{1}{1 \cdot 2 \cdot 3} f'''(x_0)(x - x_0)^3 \\ &+ \cdots \\ &+ \frac{1}{n!} f^{(n+1)}(x_0)(x - x_0)^n \\ &+ R_{n+1}(x) \end{aligned}$$

where $R_{n+1}(x)$ is called the *remainder term*. Then Taylor's theorem [61, pp. 95–96] provides that there exists some ξ between x and x_0 such that

$$R_{n+1}(x) = \frac{f^{(n+1)}(\xi)}{(n+1)!}(x - x_0)^{n+1}.$$

In particular, if $|f^{(n+1)}| \leq M$ in I, then

$$R_{n+1}(x) \leq \frac{M|x - x_0|^{n+1}}{(n+1)!}$$

which is normally small when x is close to x_0.

When $x_0 = 0$, the Taylor series reduces to what is called a *Maclaurin series* [54, p. 96].

E.4 Weierstrass Approximation Theorem

The Weierstrass approximation theorem assures us that polynomial approximation can get arbitrarily close to any continuous function as the polynomial order is increased.

Let $f(x)$ be continuous on a real interval I. Then for any $\epsilon > 0$, there exists an nth-order polynomial $P_n(f,x)$, where n depends on ϵ, such that

$$|P_n(f,x) - f(x)| < \epsilon$$

for all $x \in I$.

For a proof, see, *e.g.*, [61, pp. 146–148].

Thus, any continuous function can be approximated arbitrarily well by means of a polynomial. This does not necessarily mean that a Taylor series expansion can be used to find such a polynomial since, in particular, the function must be differentiable of all orders throughout I. Furthermore, there can be points, even in infinitely differentiable functions, about which a Taylor expansion will not yield a good approximation, as illustrated in the next section. The main point here is that, thanks to the Weierstrass approximation theorem, we know that good polynomial approximations *exist* for any continuous function.

E.5 Points of Infinite Flatness

Consider the inverted Gaussian pulse,[1]

$$f(x) \stackrel{\Delta}{=} e^{-1/x^2},$$

i.e., $f(1/x)$ is the well known Gaussian "bell curve" e^{-x^2}. Clearly, derivatives of all orders exist for all x. However, it is readily verified that *all* derivatives at $x = 0$ are zero. (It is easier to verify that all derivatives of the bell curve are zero at $x = \infty$.) Therefore, every finite-order Maclaurin series expansion of $f(x)$ is the zero function, and the Weierstrass approximation theorem cannot be fulfilled by this series.

As mentioned in §E.2, a measure of "flatness" is the number of leading zero terms in a function's Taylor expansion (not counting the first (constant) term). Thus, by this measure, the bell curve is "infinitely flat" at infinity, or, equivalently, $f(x)$ is infinitely flat at $x = 0$.

[1] Thanks to Miller Puckette for suggesting this example.

Another property of $f(x) \stackrel{\Delta}{=} e^{-\frac{1}{x^2}}$ is that it has an infinite number of "zeros" at $x = 0$. The fact that a function $g(x)$ has an infinite number of zeros at $x = x_0$ can be verified by showing

$$\lim_{x \to x_0} \frac{1}{(x - x_0)^k} g(x) = 0$$

for all $k = 1, 2, \ldots$. For $f(x)$, the existence of an infinite number of zeros at $x = 0$ is easily shown by looking at the zeros of $f(1/x)$ at $x = \infty$, *i.e.*,

$$\lim_{x \to \infty} x^k f(1/x) = \lim_{x \to \infty} x^k e^{-x^2} = 0$$

for any integer k. Thus, the faster-than-exponential decay of a Gaussian bell curve cannot be outpaced by the factor x^k, for any finite k. In other words, exponential growth or decay is faster than polynomial growth or decay. (As mentioned in §3.10 on page 25, the Taylor series expansion of the exponential function e^x is $1+x+x^2/2+x^3/3!+\ldots$—an "infinite-order" polynomial.)

The reciprocal of a function containing an infinite-order zero at $x = x_0$ has what is called an *essential singularity* at $x = x_0$ [13, p. 157], also called a *non-removable singularity*. Thus, $1/f(x) = e^{\frac{1}{x^2}}$ has an essential singularity at $x = 0$, and e^{x^2} has one at $x = \infty$.

An amazing result from the theory of complex variables [13, p. 270] is that near an essential singular point $z_0 \in \mathbf{C}$ (*i.e.*, z_0 may be a complex number), the inequality

$$|f(z) - c| < \epsilon$$

is satisfied at some point $z \neq z_0$ in *every* neighborhood of z_0, however small! In other words, the function comes arbitrarily close to every possible value in any neighborhood about an essential singular point. This result, too, is due to Weierstrass [13].

In summary, a Taylor series expansion about the point $x = x_0$ will always yield a constant approximation when the function being approximated is infinitely flat at x_0. For this reason, polynomial approximations are often applied over a restricted range of x, with constraints added to provide transitions from one interval to the next. This leads to the general subject of *splines* [78]. In particular, *cubic spline* approximations are composed of successive segments which are each third-order polynomials. In each segment, four degrees of freedom are available (the four polynomial coefficients). Two of these are usually devoted to matching the amplitude and slope of the polynomial to one side, while the other two are used to maximize some measure of fit across the segment. The points at which adjacent polynomial segments connect are called "knots", and finding optimal knot locations is usually a relatively expensive, iterative computation.

E.6 Differentiability of Audio Signals

As mentioned in §3.6 on page 23, every audio signal can be regarded as infinitely differentiable due to the finite bandwidth of human hearing. That is, given any audio signal $x(t)$, its Fourier transform is given by

$$X(\omega) \stackrel{\Delta}{=} \int_{-\infty}^{\infty} x(t)e^{-j\omega t}dt.$$

For the Fourier transform to exist, it is sufficient that $x(t)$ be absolutely integrable, *i.e.*, $\int_{-\infty}^{\infty} |x(t)|\, dt < \infty$. Clearly, all audio signals in practice are absolutely integrable. The inverse Fourier transform is then given by

$$x(t) = \frac{1}{2\pi}\int_{-\infty}^{\infty} X(\omega)e^{j\omega t}d\omega.$$

Because hearing is bandlimited to, say, 20 kHz, $x(t)$ sounds identical to the bandlimited signal

$$x_f(t) \stackrel{\Delta}{=} \frac{1}{2\pi}\int_{-\Omega}^{\Omega} X(\omega)e^{j\omega t}d\omega$$

where $\Omega \stackrel{\Delta}{=} 2\pi \cdot 20,000$. Now, taking time derivatives is simple (see also §C.1 on page 213):

$$\begin{aligned}
\frac{d}{dt}x_f(t) &= \frac{1}{2\pi}\int_{-\Omega}^{\Omega} X(\omega)(j\omega)e^{j\omega t}d\omega \\
\frac{d^2}{dt^2}x_f(t) &= \frac{1}{2\pi}\int_{-\Omega}^{\Omega} X(\omega)(j\omega)^2e^{j\omega t}d\omega \\
\vdots \quad & \quad \vdots \\
\frac{d^n}{dt^n}x_f(t) &= \frac{1}{2\pi}\int_{-\Omega}^{\Omega} X(\omega)(j\omega)^ne^{j\omega t}d\omega
\end{aligned}$$

Since the length of the integral is finite, there is no possibility that it can "blow up" due to the weighting by ω^n in the frequency domain introduced by differentiation in the time domain.

A basic Fourier property of signals and their spectra is that *a signal cannot be both time limited and frequency limited.* Therefore, by conceptually "lowpass filtering" every audio signal to reject all frequencies above 20 kHz, we implicitly make every audio signal last forever! Another way of saying this is that the "ideal lowpass filter 'rings' forever". Such fine points do not concern us in practice, but they are important for fully understanding the underlying theory. Since, in reality, signals can be said to have a true beginning and end, we must admit that all signals we really work with in practice have infinite-bandwidth. That is, when a signal is turned on or off, there is a spectral event extending all the way to infinite frequency (while "rolling off" with frequency and having a finite total energy).[2]

In summary, audio signals are perceptually equivalent to bandlimited signals, and bandlimited signals are infinitely smooth in the sense that derivatives of all orders exist at all points time $t \in (-\infty, \infty)$.

[2]One joke along these lines, due, I'm told, to Professor Bracewell at Stanford, is that "since the telephone is bandlimited to 3kHz, and since bandlimited signals cannot be time limited, it follows that one cannot hang up the telephone".

Appendix F

Logarithms and Decibels

This appendix provides an introduction to *logarithms* (real and complex) and *decibels*, a quantitative measure of sound intensity. Several specific dB scales are defined, and dynamic range considerations in audio are considered.

F.1 Logarithms

A *logarithm* $y = \log_b(x)$ is fundamentally an *exponent* y applied to a specific *base* b to yield the argument x. That is, $x = b^y$. The term "logarithm" can be abbreviated as "log". The base b is chosen to be a positive real number, and we normally only take logs of positive real numbers $x > 0$ (although it is ok to say that the log of 0 is $-\infty$). The inverse of a logarithm is called an *antilogarithm* or *antilog*; thus, x is the antilog of y in the base b.

For any positive number x, we have

$$x = b^{\log_b(x)}$$

for any valid base $b > 0$. This is just an identity arising from the definition of the logarithm, but it is sometimes useful in manipulating formulas.

When the base is not specified, it is normally assumed to be 10, *i.e.*, $\log(x) \stackrel{\Delta}{=} \log_{10}(x)$. This is the *common logarithm*.

Base 2 and base e logarithms have their own special notation:

$$\begin{aligned} \ln(x) &\stackrel{\Delta}{=} \log_e(x) \\ \lg(x) &\stackrel{\Delta}{=} \log_2(x) \end{aligned}$$

(The use of lg() for base 2 logarithms is common in computer science. In mathematics, it may denote a base 10 logarithm.) By far the most common bases are 10, e, and 2. Logs base e are called *natural logarithms*. They are "natural" in the sense that

$$\frac{d}{dx}\ln(x) = \frac{1}{x}$$

while the derivatives of logarithms to other bases are not quite so simple:

$$\frac{d}{dx}\log_b(x) = \frac{1}{x\ln(b)}$$

The inverse of the natural logarithm $y = \ln(x)$ is of course the exponential function $x = e^y$, and e^y is its own derivative.

In general, a logarithm y has an integer part and a fractional part. The integer part is called the *characteristic* of the logarithm, and the fractional part is called the *mantissa*. These terms were suggested by Henry Briggs in 1624. "Mantissa" is a Latin word meaning "addition" or "make weight"—something added to make up the weight [26].

The following Matlab code illustrates splitting a natural logarithm into its characteristic and mantissa:

```
>> x = log(3)
   x = 1.0986
>> characteristic = floor(x)
   characteristic = 1
>> mantissa = x - characteristic
   mantissa = 0.0986

>> % Now do a negative-log example
>> x = log(0.05)
   x = -2.9957
>> characteristic = floor(x)
   characteristic = -3
>> mantissa = x - characteristic
   mantissa = 0.0043
```

Logarithms were used in the days before computers to perform *multiplication of large numbers*. Since $\log(xy) = \log(x) + \log(y)$, one can look up the logs of x and y in tables of logarithms, add them together (which is easier than multiplying), and look up the antilog of the result to obtain the product xy. Log tables are still used in modern computing environments to replace expensive multiplies with less-expensive table lookups and additions. This is a classic trade-off between memory (for the log tables) and computation. Nowadays, large numbers are multiplied using FFT fast-convolution techniques.

F.1.1 Changing the Base

By definition, $x = b^{\log_b(x)}$. Taking the log base a of both sides gives

$$\log_a(x) = \log_b(x)\log_a(b)$$

which tells how to convert the base from b to a, that is, how to convert the log base b of x to the log base a of x. (Just multiply by the log base a of b.)

F.1.2 Logarithms of Negative and Imaginary Numbers

By *Euler's identity*, $e^{j\pi} = -1$, so that

$$\ln(-1) = j\pi$$

from which it follows that for any $x < 0$, $\ln(x) = j\pi + \ln(|x|)$.

Similarly, $e^{j\pi/2} = j$, so that

$$\ln(j) = j\frac{\pi}{2}$$

and for any imaginary number $z = jy$, $\ln(z) = j\pi/2 + \ln(y)$, where y is real.

Finally, from the polar representation $z = re^{j\theta}$ for complex numbers,

$$\ln(z) \stackrel{\Delta}{=} \ln(re^{j\theta}) = \ln(r) + j\theta$$

where $r > 0$ and θ are real. Thus, the log of the magnitude of a complex number behaves like the log of any positive real number, while the log of its phase term $e^{j\theta}$ extracts its phase (times j).

F.2 Decibels

A *decibel* (abbreviated dB) is defined as one tenth of a *bel*. The bel[1] is an amplitude unit defined for sound as the log (base 10) of the *intensity* relative to some *reference intensity*,[2] *i.e.*,

$$\text{Amplitude\_in\_bels} = \log_{10}\left(\frac{\text{Signal\_Intensity}}{\text{Reference\_Intensity}}\right)$$

[1] The "bel" is named after Alexander Graham Bell, the inventor of the telephone.

[2] *Intensity* is physically *power per unit area*. Bels may also be defined in terms of *energy*, or *power* which is energy per unit time. Since sound is always measured over some *area* by a microphone diaphragm, its physical power is conventionally normalized by area, giving intensity. Similarly, the *force* applied by sound to a microphone diaphragm is normalized by area to give *pressure* (force per unit area).

The choice of reference intensity (or power) defines the particular choice of *dB scale*. Signal intensity, power, and energy are always proportional to the *square* of the signal *amplitude*. Thus, we can always translate these energy-related measures into squared amplitude:

$$\text{Amplitude\_in\_bels} = \log_{10}\left(\frac{\text{Amplitude}^2}{\text{Amplitude}_{\text{ref}}^2}\right) = 2\log_{10}\left(\frac{|\text{Amplitude}|}{|\text{Amplitude}_{\text{ref}}|}\right)$$

Since there are 10 decibels to a bel, we also have

$$\begin{aligned}\text{Amplitude}_{\text{dB}} &= 20\log_{10}\left(\frac{|\text{Amplitude}|}{|\text{Amplitude}_{\text{ref}}|}\right) = 10\log_{10}\left(\frac{\text{Intensity}}{\text{Intensity}_{\text{ref}}}\right) \\ &= 10\log_{10}\left(\frac{\text{Power}}{\text{Power}_{\text{ref}}}\right) = 10\log_{10}\left(\frac{\text{Energy}}{\text{Energy}_{\text{ref}}}\right)\end{aligned}$$

A *just-noticeable difference* (JND) in amplitude level is on the order of a quarter dB. In the early days of telephony, one dB was considered a reasonable "smallest step" in amplitude, but in reality, a series of half-dB amplitude steps does not sound very smooth, while quarter-dB steps do sound pretty smooth. A typical professional audio filter-design specification for "ripple in the passband" is 0.1 dB.

F.2.1 Properties of DB Scales

In every kind of dB, a *factor of 10* in amplitude increase corresponds to a *20 dB boost* (increase by 20 dB):

$$20\log_{10}\left(\frac{10\cdot A}{A_{\text{ref}}}\right) = \underbrace{20\log_{10}(10)}_{20\text{ dB}} + 20\log_{10}\left(\frac{A}{A_{\text{ref}}}\right)$$

and $20\log_{10}(10) = 20$, of course. A function $f(x)$ which is proportional to $1/x$ is said to "fall off" (or "roll off") at the rate of 20 *dB per decade*. That is, for every factor of 10 in x (every "decade"), the amplitude drops 20 dB.

Similarly, a factor of 2 in amplitude gain corresponds to a 6 dB boost:

$$20\log_{10}\left(\frac{2\cdot A}{A_{\text{ref}}}\right) = \underbrace{20\log_{10}(2)}_{6\text{ dB}} + 20\log_{10}\left(\frac{A}{A_{\text{ref}}}\right)$$

and

$$20\log_{10}(2) = 6.0205999\ldots \approx 6\text{ dB}.$$

A function $f(x)$ which is proportional to $1/x$ is said to fall off 6 *dB per octave*. That is, for every factor of 2 in x (every "octave"), the amplitude drops close to 6 dB. Thus, 6 dB per octave is the same thing as 20 dB per decade.

A *doubling of power* corresponds to a *3 dB boost*:

$$10\log_{10}\left(\frac{2\cdot A^2}{A_{\text{ref}}^2}\right) = \underbrace{10\log_{10}(2)}_{3\text{ dB}} + 10\log_{10}\left(\frac{A^2}{A_{\text{ref}}^2}\right)$$

and

$$10\log_{10}(2) = 3.010\ldots \approx 3\text{ dB}.$$

Finally, note that the choice of *reference* merely determines a vertical offset in the dB scale:

$$20\log_{10}\left(\frac{A}{A_{\text{ref}}}\right) = 20\log_{10}(A) - \underbrace{20\log_{10}(A_{\text{ref}})}_{\text{constant offset}}$$

F.2.2 Specific DB Scales

Since we so often rescale our signals to suit various needs (avoiding overflow, reducing quantization noise, making a nicer plot, etc.), there seems to be little point in worrying about what the dB reference is—we simply choose it implicitly when we rescale to obtain signal values in the range we want to see. Nevertheless, a few specific dB scales are worth knowing about.

DBm Scale

One common dB scale in audio recording is the dBm scale in which the reference power is taken to be a milliwatt (1 mW) dissipated by a 600 Ohm resistor. (See §F.3 on page 250 for a primer on resistors, voltage, current, and power.)

DBV Scale

Another dB scale is the dBV scale which sets 0 dBV to 1 volt. Thus, a 100-volt signal is

$$20\log_{10}\left(\frac{100V}{1V}\right) = 40\text{ dBV}$$

and a 1000-volt signal is

$$20\log_{10}\left(\frac{1000V}{1V}\right) = 60\text{ dBV}$$

Note that the dBV scale is undefined for current or power, unless the voltage is assumed to be across a standard resistor value, such as 600 Ohms.

DB SPL

Sound Pressure Level (SPL) is defined using a reference which is approximately the intensity of 1000 Hz sinusoid that is just barely audible (zero "phons"). In *pressure* units:[3]

$$
\begin{aligned}
0 \text{ dB SPL} &\stackrel{\Delta}{=} 0.0002\ \mu\text{bar (micro-barometric pressure)} \\
&= 20\ \mu\text{Pa (micro-Pascals)} \\
&= 2.9 \times 10^{-9} \text{ PSI (pounds per square inch)} \\
&= 2 \times 10^{-4} \frac{\text{dynes}}{\text{cm}^2} \quad \text{(CGS units)} \\
&= 2 \times 10^{-5} \frac{\text{nt}}{\text{m}^2} \quad \text{(MKS units)}
\end{aligned}
$$

In *intensity* units:

$$I_0 = 10^{-16} \frac{\text{W}}{\text{cm}^2}$$

which corresponds to a root-mean-square (rms) pressure amplitude of 20.4 μPa, or about 20 μPa, as listed above. The *wave impedance* of air plays the role of "resistor" in relating the pressure- and intensity-based references exactly analogous to the dBm case discussed above.

Since *sound* is created by a time-varying pressure, we compute sound levels in dB-SPL by using the *average* intensity (averaged over at least one period of the lowest frequency contained in the sound).

Table F.1 gives a list of common sound levels and their dB equivalents [52]: In my experience, the "threshold of pain" is most often defined as 120 dB.

The relationship between sound amplitude and actual *loudness* is complex [73]. Loudness is a perceptual dimension while sound amplitude is physical. Since loudness sensitivity is closer to logarithmic than linear in amplitude (especially at moderate to high loudnesses), we typically use decibels to represent sound amplitude, especially in spectral displays.

The *sone amplitude scale* is defined in terms of actual loudness perception experiments [73]. At 1kHz and above, loudness perception is approximately logarithmic above 50 dB SPL or so. Below that, it tends toward being more linear.

The *phon amplitude scale* is simply the dB scale at 1kHz [73, p. 111]. At other frequencies, the amplitude in phons is defined by following the equal-loudness curve over to 1 kHz and reading off the level there in dB SPL. In other words, all pure tones have the same loudness at the same

[3]The bar was originally defined as one "atmosphere" (atmospheric pressure at sea level), but now a microbar is defined to be exactly one dyne/cm^2, where a dyne is the amount of force required to accelerate a gram by one centimeter per second squared.

Sound	dB-SPL
Jet engine at 3m	140
Threshold of pain	130
Rock concert	120
Accelerating motorcycle at 5m	110
Pneumatic hammer at 2m	100
Noisy factory	90
Vacuum cleaner	80
Busy traffic	70
Quiet restaurant	50
Residential area at night	40
Empty movie house	30
Rustling of leaves	20
Human breathing (at 3m)	10
Threshold of hearing (good ears)	0

Table F.1: Approximate dB-SPL level of common sounds. (Information from S. S. Stevens, F. Warshofsky, and the Editors of Time-Life Books, *Sound and Hearing*, Life Science Library, Time-Life Books, Alexandria, VA, 1965, p. 173.)

phon level, and 1 kHz is used to set the reference in dB SPL. Just remember that one phon is one dB-SPL at 1 kHz. Looking at the Fletcher-Munson equal-loudness curves [73, p. 124], loudness in phons can be read off along the vertical line at 1 kHz.

Classically, the *intensity level* of a sound wave is its dB SPL level, measuring the peak time-domain pressure-wave amplitude relative to 10^{-16} watts per centimeter squared (*i.e.*, there is no consideration of the frequency domain here at all).

Another classical term still encountered is the *sensation level* of pure tones: The sensation level is the number of dB SPL above the *hearing threshold* at that frequency [73, p. 110].

For further information on "doing it right," see, for example, http://www.measure.demon.co.uk/Acoustics_Software/loudness.html.

DBA (A-Weighted DB)

The so-called *A-weighted dB scale* (abbreviated *dBA*) is based on the Fletcher-Munson equal-loudness curve for an SPL of 40 phons.[4] Thus, a dBA weight-

[4]See `http://en.wikipedia.org/wiki/A-weighting` for more information, including a plot of the A weighting curve (as well as B, C, and D weightings which can be used for louder listening levels) and pointers to relevant standards.

ing assumes a fairly quiet pure tone. Despite this assumption, the dBA weighting is often used as an approximate equal loudness adjustment for measured spectra.

An analog filter transfer function that can be used to implement an approximate A-weighting is given by[5]

$$H_A(s) = \frac{k_A \cdot s^4}{(s+129.4)^2(s+676.7)(s+4636)(s+76655)^2}$$

where $k_A \approx 7.39705 \times 10^9$ normalizes the gain to unity at 1 kHz.

The *ITU-R 468 noise weighting*[6] is said to perform better for measuring noise in audio systems.

DB for Display

In practical signal processing, it is common to choose the *maximum signal magnitude* as the reference amplitude. That is, we normalize the signal so that the maximum amplitude is defined as 1, or 0 dB. This convention is also used by "sound level meters" in audio recording. When displaying magnitude spectra, the highest spectral peak is often normalized to 0 dB. We can then easily read off lower peaks as so many dB below the highest peak.

Figure F.1b shows a plot of the Fast Fourier Transform (FFT) of ten periods of a "Kaiser-windowed" sinusoid at 440 Hz. (FFT windows are introduced in §8.1.4 on page 172. The window is used to taper a finite-duration section of the signal.) Note that the peak dB magnitude has been normalized to zero, and that the plot has been clipped at -100 dB.

Below is the Matlab code for producing Fig. F.1. Note that it contains several elements (windows, zero padding, spectral interpolation) that we will not cover until later. They are included here as "forward references" in order to keep the example realistic and practical, and to give you an idea of "how far we have to go" before we know how to do practical spectrum analysis. Otherwise, the example just illustrates plotting spectra on an arbitrary dB scale between convenient limits.

```
% Practical display of the fft of a synthesized sinusoid

fs = 44100;              % Sampling rate
f = 440;                 % Sinusoidal frequency = A-440
nper = 10;               % Number of periods to synthesize
dur = nper/f;            % Duration in seconds
T = 1/fs;                % Sampling period
t = 0:T:dur;             % Discrete-time axis in seconds
```

[5] ibid.

[6] http://en.wikipedia.org/wiki/ITU-R_468_noise_weighting

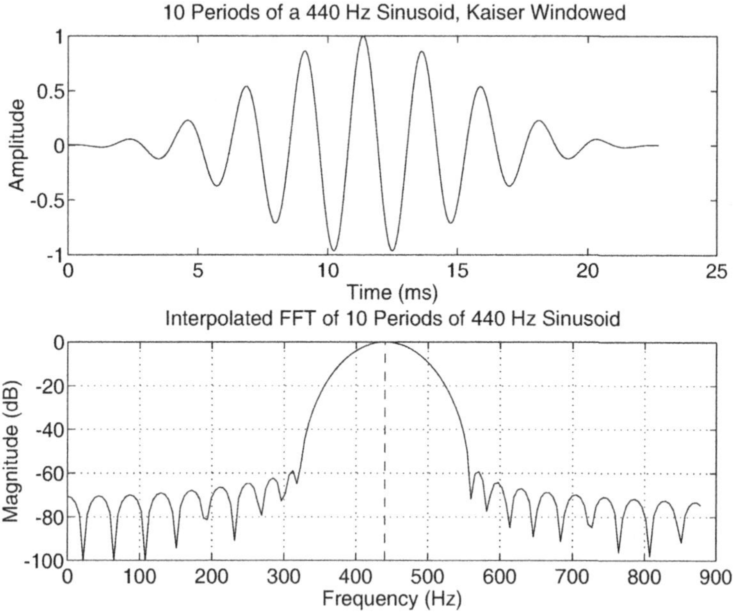

Figure F.1: Windowed sinusoid (top) and its FFT magnitude (bottom).

```
L = length(t)             % Number of samples to synthesize
ZP = 5;                   % Zero padding factor
N = 2^(nextpow2(L*ZP))  % FFT size (power of 2)

x = cos(2*pi*f*t);        % A sinusoid at A-440 ("row vector")
w = kaiser(L,8);          % An "FFT window"
xw = x .* w';             % Need to transpose w to get a row
sound(xw,fs);             % Might as well listen to it
xzp = [xw,zeros(1,N-L)];% Zero-padded FFT input buffer
X = fft(xzp);             % Interpolated spectrum of xw

Xmag = abs(X);            % Spectral magnitude
Xdb = 20*log10(Xmag);     % Spectral magnitude in dB

XdbMax = max(Xdb);        % Peak dB magnitude
Xdbn = Xdb - XdbMax;      % Normalize to 0dB peak

dBmin = -100;             % Don't show anything lower than this
Xdbp = max(Xdbn,dBmin); % Normalized, clipped, dB mag spec
fmaxp = 2*f;              % Upper frequency limit of plot, Hz
kmaxp = fmaxp*N/fs;       % Upper frequency limit of plot, bins
fp = fs*[0:kmaxp]/N;      % Frequency axis in Hz

% Ok, plot it already!

subplot(2,1,1);
plot(1000*t,xw);
xlabel('Time (ms)');
ylabel('Amplitude');
title(sprintf(['a) %d Periods of a %3.0f Hz Sinusoid, ',
               'Kaiser Windowed'],nper,f)R);

subplot(2,1,2);
plot(fp,Xdbp(1:kmaxp+1)); grid;
% Plot a dashed line where the peak should be:
  hold on; plot([440 440],[dBmin,0],'--'); hold off;
xlabel('Frequency (Hz)');
ylabel('Magnitude (dB)');
title(sprintf(['b) Interpolated FFT of %d Periods of ',...
       '%3.0f Hz Sinusoid'],nper,f));
```

The following more compact Matlab produces essentially the same plot, but without the nice physical units on the horizontal axes:

```
x = cos([0:2*pi/20:10*2*pi]); % 10 periods, 20 samples/cycle
L = length(x);
xw = x' .* kaiser(L,8);
```

```
N = 2^nextpow2(L*5);
X = fft([xw',zeros(1,N-L)]);

subplot(2,1,1); plot(xw);
xlabel('Time (samples)'); ylabel('Amplitude');
title('a) 10 Periods of a Kaiser-Windowed Sinusoid');

subplot(2,1,2); kmaxp = 2*10*5; Xl = 20*log10(abs(X(1:kmaxp+1)));
plot([10*5+1,10*5+1],[-100,0],[0:kmaxp],max(Xl-max(Xl),-100)); grid;
xlabel('Frequency (Bins)'); ylabel('Magnitude (dB)');
title('b) Interpolated FFT of 10 Periods of Sinusoid');
```

F.2.3 Dynamic Range

The *dynamic range* of a *signal processing system* can be defined as the maximum dB level sustainable without overflow (or other distortion) minus the dB level of the "noise floor".

Similarly, the dynamic range of a *signal* can be defined as its maximum decibel level minus its average "noise level" in dB. For digital signals, the limiting noise is ideally *quantization noise.*

Quantization noise is generally modeled as a uniform random variable between plus and minus half the least significant bit (since rounding to the nearest representable sample value is normally used). If q denotes the quantization interval, then the maximum quantization-error magnitude is $q/2$, and its variance ("noise power") is $\sigma_q^2 = q^2/12$ (see §G.3 for a derivation of this value).

The rms level of the quantization noise is therefore $\sigma_q = q/(2\sqrt{3}) \approx 0.3q$, or about 60% of the maximum error.

The *number system* (see Appendix G and *number of bits* chosen to represent signal samples determines their available dynamic range. Signal processing operations such as digital filtering may use the same number system as the input signal, or they may use extra bits in the computations, yielding an increased "internal dynamic range".

Since the threshold of hearing is near 0 dB SPL, and since the "threshold of pain" is often defined as 120 dB SPL, we may say that the dynamic range of human hearing is approximately 120 dB.

The dynamic range of *magnetic tape* is approximately 55 dB. To increase the dynamic range available for analog recording on magnetic tape, *companding* is often used. "Dolby A" adds approximately 10 dB to the dynamic range that will fit on magnetic tape (by compressing the signal dynamic range by 10 dB), while DBX adds 30 dB (at the cost of more "transient distortion").[7] In general, any dynamic range can be mapped to

[7]Companders (compressor-expanders) essentially "turn down" the signal gain when

any other dynamic range, subject only to noise limitations.

F.3 Voltage, Current, and Resistance

The state of an ideal resistor is completely specified by the voltage across it (call it V *volts*) and the current passing through it (I *amperes*, or simply "amps"). The ratio of voltage to current gives the value of the resistor ($V/I = R =$ resistance in Ohms). The fundamental relation between voltage and current in a resistor is called *Ohm's Law*:

$$V(t) = R \cdot I(t) \qquad \text{(Ohm's Law)}$$

where we have indicated also that the voltage and current may vary with time (while the resistor value normally does not).

The electrical *power* in *watts* dissipated by a resistor R is given by

$$\mathcal{P} = V \cdot I = \frac{V^2}{R} = R \cdot I^2$$

where V is the voltage and I is the current. Thus, volts times amps gives watts. Also, volts squared over ohms equals watts, and so on.

F.4 Exercises

1. Show that
$$\frac{d}{dx}\log_b(x) = \frac{1}{x\ln(b)}$$
where $\log_b(x)$ denotes the logarithm to the base b of x.

2. Work out the definition of logarithms using a *complex* base b.

3. Try synthesizing a sawtooth waveform which increases by 1/2 dB a few times per second, and again using 1/4 dB increments. See if you agree that quarter-dB increments are "smooth" enough for you.

---

it is "loud" and "turn up" the gain when it is "quiet". As long as the input-output curve is monotonic (such as a log characteristic), the dynamic-range compression can be undone (expanded).

Appendix G

Number Systems for Digital Audio

This appendix discusses number formats used in digital audio. They are divided into "linear" and "logarithmic." Linear number systems include binary integer fixed-point, fractional fixed-point, one's complement, and two's complement fixed-point. The μ-law format is a popular hybrid between linear and logarithmic amplitude encoding. Floating-point combines a linear mantissa with a logarithmic "exponent," and logarithmic fixed-point can be viewed as a special case of floating-point in which there is no mantissa. This appendix does not cover audio *coding* methods such as MPEG (MP3) [4, 6, 48, 5] or Linear Predictive Coding (LPC) [9, 38, 39].

G.1 Linear Number Systems

Linear number systems are used in digital audio gear such as compact disks and digital audio tapes. As such, they represent the "high end" of digital audio formats, when a sufficiently large sampling rate (*e.g.*, above 40 kHz) and number of bits per word (*e.g.*, 20), are used.

G.1.1 Pulse Code Modulation (PCM)

The "standard" number format for sampled audio signals is officially called *Pulse Code Modulation (*PCM*)*. This term simply means that each signal sample is interpreted as a "pulse" (*e.g.*, a voltage or current pulse) at a particular amplitude which is binary encoded, typically in two's complement binary fixed-point format (discussed below). When someone says they are giving you a soundfile in "raw binary format", they pretty much always mean (nowadays) 16-bit, two's-complement PCM data. Most mainstream computer soundfile formats consist of a "header" (containing the length, etc.) followed by 16-bit two's-complement PCM.

You can normally convert a soundfile from one computer's format to another by stripping off its header and prepending the header for the new machine (or simply treating it as raw binary format on the destination computer). The UNIX "cat" command can be used for this, as can the Emacs text editor (which handles binary data just fine). The only issue usually is whether the bytes have to be swapped (an issue discussed further below).

G.1.2 Binary Integer Fixed-Point Numbers

Most computer languages nowadays only offer two kinds of numbers, *floating-point* and *integer fixed-point*. On present-day computers, all numbers are encoded using *binary digits* (called "bits") which are either 1 or 0.[1] In C, C++, and Java, floating-point variables are declared as `float` (32 bits) or `double` (64 bits), while integer fixed-point variables are declared as `short int` (typically 16 bits and never less), `long int` (typically 32 bits and never less), or simply `int` (typically the same as a `long int`, but sometimes between short and long). For an 8-bit integer, one can use the `char` data type (8 bits).

Since C was designed to accommodate a wide range of hardware, including old mini-computers, some latitude was historically allowed in the choice of these bit-lengths. The `sizeof` operator is officially the "right way" for a C program to determine the number of bytes in various data types at run-time, *e.g.*, `sizeof(long)`. (The word `int` can be omitted after `short` or `long`.) Nowadays, however, `shorts` are always 16 bits (at least on all the major platforms), `ints` are 32 bits, and `longs` are typically 32 bits on

[1]Computers use bits, as opposed to the more familiar decimal digits, because they are more convenient to implement in digital hardware. For example, the *decimal numbers* 0, 1, 2, 3, 4, 5 become, in binary format, 0, 1, 10, 11, 100, 101. Each *bit* position in binary notation corresponds to a power of 2, *e.g.*, $5 = 1 \cdot 2^2 + 0 \cdot 2^1 + 1 \cdot 2^0$; while each *digit* position in decimal notation corresponds to a power of 10, *e.g.*, $123 = 1 \cdot 10^2 + 2 \cdot 10^1 + 3 \cdot 10^0$. The term "digit" comes from the same word meaning "finger." Since we have ten fingers (digits), the term "digit" technically should be associated only with decimal notation, but in practice it is used for others as well. Other popular number systems in computers include *octal* which is base 8 (rarely seen any more, but still specifiable in any C/C++ program by using a leading zero, *e.g.*, $0755 = 7 \cdot 8^2 + 5 \cdot 8^1 + 5 \cdot 8^0 = 493$ decimal = 111,101,101 binary), and *hexadecimal* (or simply "*hex*") which is base 16 and which employs the letters A through F to yield 16 digits (specifiable in C/C++ by starting the number with "0x", *e.g.*, 0x1ED $= 1 \cdot 16^2 + 15 \cdot 16^1 + 14 \cdot 16^0 = 493$ decimal = 1,1110,1101 binary). Note, however, that the representation within the computer is still always binary; octal and hex are simply convenient *groupings* of bits into sets of three bits (octal) or four bits (hex).

32-bit computers and 64 bits on 64-bit computers (although some C/C++ compilers use `long long int` to declare 64-bit ints). Table G.1 gives the lengths currently used by GNU C/C++ compilers (usually called "`gcc`" or "`cc`") on 64-bit processors.[2]

Type	Bytes	Notes
`char`	1	
`short`	2	
`int`	4	
`long`	8	(4 bytes on 32-bit machines)
`long long`	8	(may become 16 bytes)
`type *`	8	(any pointer)
`float`	4	
`double`	8	
`long double`	8	(may become 10 bytes)
`size_t`	8	(type of sizeof())
T* - T*	8	(pointer arithmetic)

Table G.1: Byte sizes of GNU C/C++ data types for 64-bit machines.

Java, which is designed to be platform independent, defines a `long int` as *equivalent in precision to* 64 bits, an `int` as 32 bits, a `short int` as 16 bits, and additionally a `byte int` as an 8-bit int. Similarly, the "Structured Audio Orchestra Language" (SAOL[3]) (pronounced "sail")—the sound-synthesis component of the new MPEG-4 audio compression standard—requires only that the underlying number system be at least as accurate as 32-bit `floats`. All `ints` discussed thus far are *signed* integer formats. C and C++ also support *unsigned* versions of all `int` types, and they range from 0 to $2^N - 1$ instead of -2^{N-1} to $2^{N-1} - 1$, where N is the number of bits. Finally, an `unsigned char` is often used for integers that only range between 0 and 255.

One's Complement Fixed-Point Format

One's Complement is a particular assignment of bit patterns to numbers. For example, in the case of 3-bit binary numbers, we have the assignments shown in Table G.2.

In general, N-bit numbers are assigned to binary counter values in the "obvious way" as integers from 0 to $2^{N-1} - 1$, and then the negative

[2]This information is subject to change without notice. Check your local compiler documentation.

[3]`http://sound.media.mit.edu/mpeg4`

Binary	Decimal
000	0
001	1
010	2
011	3
100	-3
101	-2
110	-1
111	-0

Table G.2: Three-bit one's-complement binary fixed-point numbers.

numbers are assigned in reverse order, as shown in the example.

The term "one's complement" refers to the fact that negating a number in this format is accomplished by simply *complementing* the bit pattern (inverting each bit).

Note that there are two representations for zero (all 0s and all 1s). This is inconvenient when testing if a number is equal to zero. For this reason, one's complement is generally not used.

Two's Complement Fixed-Point Format

In *two's complement*, numbers are negated by complementing the bit pattern and *adding 1*, with overflow ignored. From 0 to $2^{N-1} - 1$, positive numbers are assigned to binary values exactly as in one's complement. The remaining assignments (for the negative numbers) can be carried out using the two's complement negation rule. Regenerating the $N = 3$ example in this way gives Table G.3.

Binary	Decimal
000	0
001	1
010	2
011	3
100	-4
101	-3
110	-2
111	-1

Table G.3: Three-bit two's-complement binary fixed-point numbers.

Note that according to our negation rule, $-(-4) = -4$. Logically, what has happened is that the result has "overflowed" and "wrapped around" back to itself. Note that $3 + 1 = -4$ also. In other words, if you compute 4 somehow, since there is no bit-pattern assigned to 4, you get -4, because -4 is assigned the bit pattern that would be assigned to 4 if N were larger. Note that numerical overflows naturally result in "wrap around" from positive to negative numbers (or from negative numbers to positive numbers). Computers normally "trap" overflows as an "exception." The exceptions are usually handled by a software "interrupt handler," and this can greatly slow down the processing by the computer (one numerical calculation is being replaced by a rather sizable program).

Note that *temporary overflows* are ok in two's complement; that is, if you add 1 to 3 to get -4, adding -1 to -4 will give 3 again. This is why two's complement is a nice choice: it can be thought of as placing all the numbers on a "ring," allowing temporary overflows of intermediate results in a long string of additions and/or subtractions. All that matters is that the final sum lie within the supported dynamic range.

Computers designed with signal processing in mind (such as so-called "Digital Signal Processing (DSP) chips") generally just do the best they can without generating exceptions. For example, overflows quietly "saturate" instead of "wrapping around" (the hardware simply replaces the overflow result with the maximum positive or negative number, as appropriate, and goes on). Since the programmer may wish to know that an overflow has occurred, the first occurrence may set an "overflow indication" bit which can be manually cleared. The overflow bit in this case just says an overflow happened sometime since it was last checked.

Two's-Complement, Integer Fixed-Point Numbers

Let N denote the number of bits. Then the value of a two's complement integer fixed-point number can be expressed in terms of its bits $\{b_i\}_{i=0}^{N-1}$ as

$$x = -b_0 \cdot 2^{N-1} + \sum_{i=1}^{N-1} b_i \cdot 2^{N-1-i}, \quad b_i \in \{0, 1\} \tag{G.1}$$

We visualize the *binary word* containing these bits as

$$x = [b_0\, b_1\, \cdots\, b_{N-1}]$$

Each bit b_i is of course either 0 or 1. Check that the $N = 3$ case in Table G.3 on page 254 is computed correctly using this formula. As an example, the number 3 is expressed as

$$3 = [011] = -0 \cdot 4 + 1 \cdot 2 + 1 \cdot 1$$

while the number -3 is expressed as

$$-3 = [101] = -1 \cdot 4 + 0 \cdot 2 + 1 \cdot 1$$

and so on.

The most-significant bit in the word, b_0, can be interpreted as the "sign bit". If b_0 is "on", the number is negative. If it is "off", the number is either zero or positive.

The least-significant bit is b_{N-1}. "Turning on" that bit adds 1 to the number, and there are no fractions allowed.

The largest positive number is when all bits are on except b_0, in which case $x = 2^{N-1} - 1$. The largest (in magnitude) negative number is $10 \cdots 0$, *i.e.*, $b_0 = 1$ and $b_i = 0$ for all $i > 0$. Table G.4 shows some of the most common cases.

N	$x_{\min}$	$x_{\max}$
8	-128	127
16	-32768	32767
24	-8,388,608	8,388,607
32	-2,147,483,648	2,147,483,647

Table G.4: Numerical range limits in N-bit two's-complement.

G.1.3 Fractional Binary Fixed-Point Numbers

In "DSP chips" (microprocessors explicitly designed for digital signal processing applications), the most commonly used fixed-point format is *fractional fixed point*, also in two's complement.

Quite simply, fractional fixed-point numbers are obtained from integer fixed-point numbers by dividing them by 2^{N-1}. Thus, the only difference is a scaling of the assigned numbers. In the $N = 3$ case, we have the correspondences shown in Table G.5.

G.1.4 How Many Bits are Enough for Digital Audio?

Armed with the above knowledge, we can visit the question "how many bits are enough" for digital audio. Since the threshold of hearing is around 0 db SPL, the "threshold of pain" is around 120 dB SPL, and each bit in a linear PCM format is worth about $20\log_{10}(2) \approx 6$ dB of dynamic range, we find that we need $120/6 = 20$ bits to represent the full dynamic range of audio in a linear fixed-point format. This is a simplistic analysis because it is not quite right to equate the least-significant bit with the threshold of hearing;

Binary	Decimal	
000	0	(0/4)
001	0.25	(1/4)
010	0.5	(2/4)
011	0.75	(3/4)
100	-1	(-4/4)
101	-0.75	(-3/4)
110	-0.5	(-2/4)
111	-0.25	(-1/4)

Table G.5: Three-bit fractional fixed-point numbers.

instead, we would like to adjust the *quantization noise floor* to just below the threshold of hearing. Since the threshold of hearing is non-uniform, we would also prefer a *shaped* quantization noise floor (a feat that can be accomplished using *filtered error feedback*[4].) Nevertheless, the simplistic result gives an answer similar to the more careful analysis, and 20 bits is a good number. However, this still does not provide for *headroom* needed in a digital recording scenario. We also need both headroom and *guard bits* on the lower end when we plan to carry out a lot of signal processing operations, especially digital filtering. As an example, a 1024-point FFT (Fast Fourier Transform) can give amplitudes 1024 times the input amplitude (such as in the case of a constant "dc" input signal), thus requiring 10 headroom bits. In general, 24 fixed-point bits are pretty reasonable to work with, although you still have to scale very carefully, and 32 bits are preferable.

G.1.5 When Do We Have to Swap Bytes?

When moving a soundfile from one computer to another, such as from a "PC" to a "Mac" (Intel processor to Motorola processor), the bytes in each sound sample have to be *swapped*. This is because Motorola processors are *big endian* (bytes are numbered from most-significant to least-significant in a multi-byte word) while Intel processors are *little endian* (bytes are numbered from least-significant to most-significant).[5] Any Mac program that supports a soundfile format native to PCs (such as `.wav` files) will

[4]Normally, quantization error is computed as $e(n) = x(n) - \hat{x}(n)$, where $x(n)$ is the signal being quantized, and $\hat{x}(n) = Q[x(n)]$ is the quantized value, obtained by rounding to the nearest representable amplitude. Filtered error feedback uses instead the formula $\hat{x}(n) = Q[x(n)+\mathcal{L}\{e(n-1)\}]$, where $\mathcal{L}\{\ \}$ denotes a filtering operation which "shapes" the quantization noise spectrum. An excellent article on the use of round-off error feedback in audio digital filters is [15].

[5]Remember that byte addresses in a big endian word start at the big end of the word, while in a little endian architecture, they start at the little end of the word.

swap the bytes for you. You only have to worry about swapping the bytes yourself when reading `raw binary` soundfiles from a foreign computer, or when digging the sound samples out of an "unsupported" soundfile format yourself.

Since soundfiles typically contain 16 bit samples (not for any good reason, as we now know), there are only two bytes in each audio sample. Let L denote the least-significant byte, and M the most-significant byte. Then a 16-bit word is most naturally written $[M, L] = M \cdot 256 + L$, *i.e.*, the most-significant byte is most naturally written to the left of the least-significant byte, analogous to the way we write binary or decimal integers. This "most natural" ordering is used as the byte-address ordering in big-endian processors:

M,L, M,L, M,L, ..., M,L (Big Endian)

Little-endian machines, on the other hand, store bytes in the order

L,M, L,M, L,M, ..., L,M. (Little Endian)

These orderings are preserved when the sound data are written to a disk file.

Since a byte (eight bits) is the smallest addressable unit in modern day processor families, we don't have to additionally worry about reversing the *bits* in each byte. Bits are not given explicit "addresses" in memory. They are extracted by means other than simple addressing (such as masking and shifting operations, table look-up, or using specialized processor instructions).

Table G.6 lists popular present-day processors and their "endianness":[6]

Processor Family	Endian
Pentium (Intel)	Little
Alpha (DEC/Compaq)	Little
680x0 (Motorola)	Big
PowerPC (Motorola & IBM)	Big
SPARC (Sun)	Big
MIPS (SGI)	Big

Table G.6: Byte ordering in the major computing platforms.

When compiling C or C++ programs under UNIX, there may be a macro constant called `BYTE_ORDER` in the header file `endian.h` or `bytesex.h`.

[6] Thanks to Bill Schottstaedt for help with this table.

In other cases, there may be macros such as `__INTEL__`, `__LITTLE_ENDIAN__`, `__BIG_ENDIAN__`, or the like, which can be detected at compile time using `#ifdef`.

G.2 Logarithmic Number Systems for Audio

Since hearing is approximately logarithmic, it makes sense to represent sound samples in a logarithmic or semi-logarithmic number format. Floating-point numbers in a computer are partially logarithmic (the exponent part), and one can even use an entirely logarithmic fixed-point number system. The μ-law amplitude-encoding format is linear at small amplitudes and becomes logarithmic at large amplitudes. This section discusses these formats.

G.2.1 Floating-Point Numbers

Floating-point numbers consist of an "exponent," "significand", and "sign bit". For a negative number, we may set the sign bit of the floating-point word and negate the number to be encoded, leaving only nonnegative numbers to be considered. Zero is represented by all zeros, so now we need only consider positive numbers.

The basic idea of floating point *encoding* of a binary number is to *normalize* the number by *shifting* the bits either left or right until the shifted result lies between 1/2 and 1. (A left-shift by one place in a binary word corresponds to multiplying by 2, while a right-shift one place corresponds to dividing by 2.) The number of bit-positions shifted to normalize the number can be recorded as a signed integer. The negative of this integer (*i.e.*, the shift required to recover the original number) is defined as the *exponent* of the floating-point encoding. The normalized number between 1/2 and 1 is called the *significand*, so called because it holds all the "significant bits" of the number.

Floating point notation is exactly analogous to "scientific notation" for decimal numbers, *e.g.*, 1.2345×10^{-9}; the number of significant digits, 5 in this example, is determined by counting digits in the "significand" 1.2345, while the "order of magnitude" is determined by the power of 10 (-9 in this case). In floating-point numbers, the significand is stored in fractional two's-complement binary format, and the exponent is stored as a binary integer.

Since the significand lies in the interval $[1/2, 1)$,[7] its most significant bit is always a 1, so it is not actually stored in the computer word, giving one more significant bit of precision.

[7] The notation $[a, b)$ denotes a *half-open interval* which includes a but not b.

Let's now restate the above a little more precisely. Let $x > 0$ denote a number to be encoded in floating-point, and let $\tilde{x} = x \cdot 2^{-E}$ denote the normalized value obtained by shifting x either E bits to the right (if $E > 0$), or $|E|$ bits to the left (if $E < 0$). Then we have $1/2 \leq \tilde{x} < 1$, and $x = \tilde{x} \cdot 2^{E}$. The *significand* M of the floating-point representation for x is defined as the binary encoding of $\tilde{x}$.[8] It is often the case that $\tilde{x}$ requires more bits than are available for exact encoding. Therefore, the significand is typically *rounded* (or truncated) to the value closest to $\tilde{x}$. Given N_M bits for the significand, the encoding of $\tilde{x}$ can be computed by multiplying it by 2^{N_M} (left-shifting it N_M bits), *rounding* to the nearest integer (or *truncating* toward minus infinity—as implemented by the `floor()` function), and encoding the N_M-bit result as a binary (signed) integer.

As a final practical note, exponents in floating-point formats may have a *bias*. That is, instead of storing E as a binary integer, you may find a binary encoding of $E - B$ where B is the bias.[9]

These days, floating-point formats generally follow the IEEE standards set out for them. A single-precision floating point word is 32 bits (four bytes) long, consisting of 1 sign bit, 8 exponent bits, and 23 significand bits, normally laid out as

S EEEEEEEE MMMMMMMMMMMMMMMMMMMMMMM

where S denotes the sign bit, E an exponent bit, and M a significand bit. Note that in this layout, ordinary integer comparison can be used in the hardware.

A double-precision floating point word is 64 bits (eight bytes) long, consisting of 1 sign bit, 11 exponent bits, and 52 significand bits. In the Intel Pentium processor, there is also an *extended precision* format, used for intermediate results, which is 80 bits (ten bytes) containing 1 sign bit, 15 exponent bits, and 64 significand bits. In Intel processors, the exponent bias is 127 for single-precision floating-point, 1023 for double-precision, and 16383 for extended-precision. The single and double precision formats have a "hidden" significand bit, while the extended precision format does not. Thus, the most significant significand bit is always set in extended precision.

The MPEG-4 audio compression standard (which supports compression using music synthesis algorithms) specifies that the numerical calcula-

[8]Another term commonly heard for "significand" is "mantissa." However, this use of the term "mantissa" is not the same as its previous definition as the fractional part of a logarithm. We will therefore use only the term "significand" to avoid confusion.

[9]By choosing the bias equal to half the numerical dynamic range of E (thus effectively inverting the sign bit of the exponent), it becomes easier to compare two floating-point numbers in hardware: the entire floating-point word can be treated by the hardware as one giant integer for numerical comparison purposes. This works because negative exponents correspond to floating-point numbers less than 1 in magnitude, while positive exponents correspond to floating-point numbers greater than 1 in magnitude.

tions in any MPEG-4 audio decoder should be at least as accurate as 32-bit single-precision floating point.

G.2.2 Logarithmic Fixed-Point Numbers

In some situations it makes sense to use *logarithmic fixed-point.* This number format can be regarded as a floating-point format consisting of an exponent and no explicit significand. However, the exponent is not interpreted as an integer as it is in floating point. Instead, it has a fractional part which is a true *mantissa.* (The integer part is then the "characteristic" of the logarithm.) In other words, a logarithmic fixed-point number is a binary encoding of the log-base-2 of the signal-sample magnitude. The sign bit is of course separate.

An example 16-bit logarithmic fixed-point number format suitable for digital audio consists of one sign bit, a 5-bit characteristic, and a 10-bit mantissa:

S CCCCC MMMMMMMMMM

The 5-bit characteristic gives a dynamic range of about 6 dB $\times 2^5 = 192$ dB. This is an excellent dynamic range for digital audio. (While 120 dB would seem to be enough for audio, consider that when digitally modeling a brass musical instrument, say, the internal air pressure near the "virtual mouthpiece" can be far higher than what actually reaches the ears in the audience.)

A nice property of logarithmic fixed-point numbers is that multiplies simply become additions and divisions become subtractions. The hard elementary operation are now addition and subtraction, and these are normally done using table lookups to keep them simple.

One "catch" when working with logarithmic fixed-point numbers is that you can't let "dc" build up. A wandering dc component will cause the quantization to be coarse even for low-level "ac" signals. It's a good idea to make sure dc is always filtered out in logarithmic fixed-point.

G.2.3 Mu-Law Coding

Digital telephone CODECs[10] have historically used (for land-line switching networks) a simple 8-bit format called μ*-law* (or simply "mu-law") that compresses large amplitudes in a manner loosely corresponding to human loudness perception.

Given an input sample $x(n)$ represented in some internal format, such

[10]CODEC is an acronym for "COder/DECoder".

as a `short`, it is converted to 8-bit mu-law format by the formula [56]

$$\hat{x}_\mu \triangleq Q_\mu \left[\log_2 \left(1 + \mu \left|x(n)\right|\right)\right]$$

where $Q_\mu[]$ is a *quantizer* which produces a kind of logarithmic fixed-point number with a 3-bit characteristic and a 4-bit mantissa, using a small table lookup for the mantissa.

As we all know from talking on the telephone, mu-law sounds really quite good for voice, at least as far as *intelligibility* is concerned. However, because the telephone bandwidth is only around 3 kHz (nominally 200–3200 Hz), there is very little "bass" and no "highs" in the spectrum above 4 kHz. This works out fine for intelligibility of voice because the first three *formants* (envelope peaks) in typical speech spectra occur in this range, and also because the difference in spectral shape (particularly at high frequencies) between consonants such as "sss", "shshsh", "fff", "ththth", etc., are sufficiently preserved in this range. As a result of the narrow bandwidth provided for speech, it is sampled at only 8 kHz in standard CODEC chips.

For "wideband audio", we like to see sampling rates at least as high as 44.1 kHz, and the latest systems are moving to 96 kHz (mainly because oversampling simplifies signal processing requirements in various areas, not because we can actually hear anything above 20 kHz). In addition, we like the low end to extend at least down to 20 Hz or so. (The lowest note on a normally tuned bass guitar is E1 = 41.2 Hz. The lowest note on a grand piano is A0 = 27.5 Hz.)

G.3 Round-Off Error Variance

This appendix shows how to derive that the noise power of amplitude quantization error is $q^2/12$, where q is the quantization step size. This is an example of a topic in *statistical signal processing*, which is beyond the scope of this book. (Some good textbooks in this area include [25, 49, 32, 31, 62, 30].) However, since the main result is so useful in practice, it is derived below anyway, with needed definitions given along the way. The interested reader is encouraged to explore one or more of the above-cited references in statistical signal processing.[11]

[11] The first by Gray and Davisson is available free online. The second by Papoulis is a classic textbook. The two volumes by Kay provide perhaps the most comprehensive coverage of the field. The volumes by Sharf and Kailath represent material used for many years in the authors' respective graduate level courses in statistical signal processing. All of the cited authors are well known researchers and professors in the field. It should also perhaps be noted that Book IV [67] in the music signal processing book series (of which this is Book I) contains a fair amount of introductory material in this area.

Each round-off error in quantization noise $e(n)$ is modeled as a uniform *random variable* between $-q/2$ and $q/2$. It therefore has the following *probability density function* (pdf) [49]:[12]

$$p_e(x) = \begin{cases} \frac{1}{q}, & |x| \le \frac{q}{2} \\ 0, & |x| > \frac{q}{2} \end{cases}$$

Thus, the *probability* that a given roundoff error $e(n)$ lies in the interval $[x_1, x_2]$ is given by

$$\int_{x_1}^{x_2} p_e(x)dx = \frac{x_2 - x_1}{q}$$

assuming of course that x_1 and x_2 lie in the allowed range $[-q/2, q/2]$. We might loosely refer to $p_e(x)$ as a *probability distribution*, but technically it is a probability *density* function, and to obtain probabilities, we have to integrate over one or more intervals, as above. We use probability *distributions* for variables which take on *discrete* values (such as dice), and we use probability *densities* for variables which take on *continuous* values (such as round-off errors).

The *mean* of a random variable is defined as

$$\mu_e \stackrel{\Delta}{=} \int_{-\infty}^{\infty} xp_e(x)dx = 0.$$

In our case, the mean is zero because we are assuming the use of *rounding* (as opposed to truncation, etc.).

The *mean of a signal* $e(n)$ is the same thing as the *expected value* of $e(n)$, which we write as $\mathcal{E}\{e(n)\}$. In general, the expected value of *any* function $f(v)$ of a random variable v is given by

$$\mathcal{E}\{f(v)\} \stackrel{\Delta}{=} \int_{-\infty}^{\infty} f(x)p_v(x)dx.$$

Since the quantization-noise signal $e(n)$ is modeled as a series of independent, identically distributed (iid) random variables, we can *estimate* the mean by *averaging* the signal over time. Such an estimate is called a *sample mean*.

Probability distributions are often characterized by their *moments*. The nth moment of the pdf $p(x)$ is defined as

$$\int_{-\infty}^{\infty} x^n p(x)dx.$$

---

[12] `http://en.wikipedia.org/wiki/Probability_density_function`

Thus, the mean $\mu_x = \mathcal{E}\{e(n)\}$ is the *first* moment of the pdf. The second moment is simply the expected value of the random variable squared, *i.e.*, $\mathcal{E}\{e^2(n)\}$.

The *variance* of a random variable $e(n)$ is defined as the *second central moment* of the pdf:

$$\sigma_e^2 \triangleq \mathcal{E}\{[e(n) - \mu_e]^2\} = \int_{-\infty}^{\infty} (x - \mu_e)^2 p_e(x) dx$$

"Central" just means that the moment is evaluated after subtracting out the *mean*, that is, looking at $e(n) - \mu_e$ instead of $e(n)$. In the case of round-off errors, the mean is zero, so subtracting out the mean has no effect. Plugging in the constant pdf for our random variable $e(n)$ which we assume is uniformly distributed on $[-q/2, q/2]$, we obtain the variance

$$\sigma_e^2 = \int_{-q/2}^{q/2} x^2 \frac{1}{q} dx = \frac{1}{q} \frac{1}{3} x^3 \bigg|_{-q/2}^{q/2} = \frac{q^2}{12}.$$

Note that the variance of $e(n)$ can be estimated by averaging $e^2(n)$ over time, that is, by computing the *mean square*. Such an estimate is called the *sample variance*. For sampled physical processes, the sample variance is proportional to the *average power* in the signal. Finally, the square root of the sample variance (the *rms level*) is sometimes called the *standard deviation* of the signal, but this term is only precise when the random variable has a Gaussian pdf.

Appendix H

Matrices

A *matrix* is defined as a rectangular array of numbers, *e.g.*,

$$\mathbf{A} = \begin{bmatrix} a & b \\ c & d \end{bmatrix}$$

which is a 2×2 ("two by two") matrix. A general matrix may be $M \times N$, where M is the number of *rows*, and N is the number of *columns* of the matrix. For example, the general 3×2 matrix is

$$\begin{bmatrix} a & b \\ c & d \\ e & f \end{bmatrix}.$$

Either square brackets or large parentheses may be used to delimit the matrix. The (i, j)th element[1] of a matrix $\mathbf{A}$ may be denoted by $\mathbf{A}[i, j]$, $\mathbf{A}(i, j)$, or $\mathbf{A}_{ij}$. For example, $\mathbf{A}[1, 2] = b$ in the above two examples. The rows and columns of matrices are normally numbered from 1 instead of from 0; thus, $1 \leq i \leq M$ and $1 \leq j \leq N$. When $N = M$, the matrix is said to be *square*.

The *transpose* of a real matrix $\mathbf{A} \in \mathbf{R}^{M \times N}$ is denoted by $\mathbf{A}^{\mathrm{T}}$ and is defined by

$$\mathbf{A}^{\mathrm{T}}[i, j] \triangleq \mathbf{A}[j, i].$$

While $\mathbf{A}$ is $M \times N$, its transpose is $N \times M$. We may say that the "rows and columns are interchanged" by the transpose operation, and transposition can be visualized as "flipping" the matrix about its main diagonal. For

[1] We are now using j as an integer counter, not as $\sqrt{-1}$. This is standard notational practice.

example,

$$\begin{bmatrix} a & b \\ c & d \\ e & f \end{bmatrix}^{\mathrm{T}} = \begin{bmatrix} a & c & e \\ b & d & f \end{bmatrix}.$$

A *complex matrix* $\mathbf{A} \in \mathbf{C}^{M \times N}$, is simply a matrix containing complex numbers. The *transpose* of a complex matrix is normally defined to include *conjugation*. The conjugating transpose operation is called the *Hermitian transpose*. To avoid confusion, in this tutorial, $\mathbf{A}^{\mathrm{T}}$ and the word "transpose" will always denote transposition *without* conjugation, while conjugating transposition will be denoted by A^* and be called the "Hermitian transpose" or the "conjugate transpose." Thus,

$$A^*[i,j] \triangleq \overline{\mathbf{A}[j,i]}.$$

Example: The Hermitian transpose of the general 3×2 matrix is

$$\begin{bmatrix} a & b \\ c & d \\ e & f \end{bmatrix}^{*} = \begin{bmatrix} \overline{a} & \overline{c} & \overline{e} \\ \overline{b} & \overline{d} & \overline{f} \end{bmatrix}.$$

A *column vector*, *e.g.*,

$$\underline{x} = \begin{bmatrix} x_0 \\ x_1 \end{bmatrix}$$

is the special case of an $M \times 1$ matrix, while a *row vector*, *e.g.*,

$$\underline{x}^{\mathrm{T}} = [\, x_0 \; x_1 \,]$$

(as we have been using) is a $1 \times N$ matrix. It is often helpful to adopt the convention that all vectors written without the transpose notation are *column vectors*, so that all row vectors require the transpose notation, as in the equation above.

H.1 Matrix Multiplication

Let $\mathbf{A}^{\mathrm{T}}$ be a general $M \times L$ matrix and let $\mathbf{B}$ denote a general $L \times N$ matrix. Denote the matrix product by $\mathbf{C} = \mathbf{A}^{\mathrm{T}}\mathbf{B}$. Then *matrix multiplication* is carried out by computing the *inner product* of every row of $\mathbf{A}^{\mathrm{T}}$ with every column of $\mathbf{B}$. Let the ith row of $\mathbf{A}^{\mathrm{T}}$ be denoted by $\underline{a}_i^{\mathrm{T}}$, $i = 1, 2, \ldots, M$, and the jth column of $\mathbf{B}$ by $\underline{b}_j$, $j = 1, 2, \ldots, N$. Then the matrix product

$\mathbf{C} = \mathbf{A}^{\mathrm{T}}\mathbf{B}$ is defined as

$$\mathbf{C} = \mathbf{A}^{\mathrm{T}}\mathbf{B} = \begin{bmatrix} <\underline{a}_1^{\mathrm{T}},\underline{b}_1> & <\underline{a}_1^{\mathrm{T}},\underline{b}_2> & \cdots & <\underline{a}_1^{\mathrm{T}},\underline{b}_N> \\ <\underline{a}_2^{\mathrm{T}},\underline{b}_1> & <\underline{a}_2^{\mathrm{T}},\underline{b}_2> & \cdots & <\underline{a}_2^{\mathrm{T}},\underline{b}_N> \\ \vdots & \vdots & \cdots & \vdots \\ <\underline{a}_M^{\mathrm{T}},\underline{b}_1> & <\underline{a}_M^{\mathrm{T}},\underline{b}_2> & \cdots & <\underline{a}_M^{\mathrm{T}},\underline{b}_N> \end{bmatrix}.$$

This definition can be extended to *complex* matrices by using a definition of inner product which does not conjugate its second argument.[2]

Examples:

$$\begin{bmatrix} a & b \\ c & d \\ e & f \end{bmatrix} \cdot \begin{bmatrix} \alpha & \beta \\ \gamma & \delta \end{bmatrix} = \begin{bmatrix} a\alpha + b\gamma & a\beta + b\delta \\ c\alpha + d\gamma & c\beta + d\delta \\ e\alpha + f\gamma & e\beta + f\delta \end{bmatrix}$$

$$\begin{bmatrix} \alpha & \beta \\ \gamma & \delta \end{bmatrix} \cdot \begin{bmatrix} a & c & e \\ b & d & f \end{bmatrix} = \begin{bmatrix} \alpha a + \beta b & \alpha c + \beta d & \alpha e + \beta f \\ \gamma a + \delta b & \gamma c + \delta d & \gamma e + \delta f \end{bmatrix}$$

$$\begin{bmatrix} \alpha \\ \beta \end{bmatrix} \cdot \begin{bmatrix} a & b & c \end{bmatrix} = \begin{bmatrix} \alpha a & \alpha b & \alpha c \\ \beta a & \beta b & \beta c \end{bmatrix}$$

$$\begin{bmatrix} a & b & c \end{bmatrix} \cdot \begin{bmatrix} \alpha \\ \beta \\ \gamma \end{bmatrix} = a\alpha + b\beta + c\gamma$$

An $M \times L$ matrix $\mathbf{A}$ can be multiplied on the *right* by an $L \times N$ matrix, where N is any positive integer. An $L \times N$ matrix $\mathbf{A}$ can be multiplied on the *left* by a $M \times L$ matrix, where M is any positive integer. Thus, the number of columns in the matrix on the left must equal the number of rows in the matrix on the right.

Matrix multiplication is *non-commutative*, in general. That is, normally $\mathbf{A}\,\mathbf{B} \neq \mathbf{B}\,\mathbf{A}$ even when both products are defined (such as when the matrices are square.)

The *transpose of a matrix product* is the product of the transposes in *reverse order*:

$$(\mathbf{AB})^{\mathrm{T}} = \mathbf{B}^{\mathrm{T}}\mathbf{A}^{\mathrm{T}}$$

---

[2] Alternatively, it can be extended to the complex case by writing $\mathbf{A}^*\mathbf{B} \stackrel{\Delta}{=} [\ldots < b_j, a_i^* > \ldots]$, so that $\mathbf{A}^*$ includes a conjugation of the elements of $\mathbf{A}$. This difficulty arises from the fact that matrix multiplication is really defined without consideration of conjugation or transposition at all, making it unwieldy to express in terms of inner products in the complex case, even though that is perhaps the most fundamental interpretation of a matrix multiply.

The *identity matrix* is denoted by $\mathbf{I}$ and is defined as

$$\mathbf{I} \triangleq \begin{bmatrix} 1 & 0 & 0 & \cdots & 0 \\ 0 & 1 & 0 & \cdots & 0 \\ 0 & 0 & 1 & \cdots & 0 \\ \vdots & \vdots & \vdots & \cdots & \vdots \\ 0 & 0 & 0 & \cdots & 1 \end{bmatrix}$$

Identity matrices are always *square*. The $N \times N$ identity matrix $\mathbf{I}$, sometimes denoted as $\mathbf{I}_N$, satisfies $\mathbf{A} \cdot \mathbf{I}_N = \mathbf{A}$ for every $M \times N$ matrix $\mathbf{A}$. Similarly, $\mathbf{I}_M \cdot \mathbf{A} = \mathbf{A}$, for every $M \times N$ matrix $\mathbf{A}$.

As a special case, a matrix $\mathbf{A}^{\mathrm{T}}$ times a vector $\underline{x}$ produces a new vector $\underline{y} = \mathbf{A}^{\mathrm{T}}\underline{x}$ which consists of the inner product of every row of $\mathbf{A}^{\mathrm{T}}$ with $\underline{x}$

$$\mathbf{A}^{\mathrm{T}}\underline{x} = \begin{bmatrix} < \underline{a}_1^{\mathrm{T}}, \underline{x} > \\ < \underline{a}_2^{\mathrm{T}}, \underline{x} > \\ \vdots \\ < \underline{a}_M^{\mathrm{T}}, \underline{x} > \end{bmatrix}.$$

A matrix $\mathbf{A}^{\mathrm{T}}$ times a vector $\underline{x}$ defines a *linear transformation* of $\underline{x}$. In fact, every linear function of a vector $\underline{x}$ can be expressed as a matrix multiply. In particular, every linear *filtering* operation can be expressed as a matrix multiply applied to the input signal. As a special case, every linear, time-invariant (LTI) filtering operation can be expressed as a matrix multiply in which the matrix is *Toeplitz*, *i.e.*, $\mathbf{A}^{\mathrm{T}}[i,j] = \mathbf{A}^{\mathrm{T}}[i-j]$ (constant along *diagonals*).

As a further special case, a row vector on the left may be multiplied by a column vector on the right to form a *single inner product*:

$$\underline{y}^*\underline{x} = \langle \underline{x}, \underline{y} \rangle$$

H.2 Solving Linear Equations Using Matrices

Consider the linear system of equations

$$\begin{aligned} ax_1 + bx_2 &= c \\ dx_1 + ex_2 &= f \end{aligned}$$

in matrix form:

$$\begin{bmatrix} a & b \\ d & e \end{bmatrix} \begin{bmatrix} x_1 \\ x_2 \end{bmatrix} = \begin{bmatrix} c \\ f \end{bmatrix}$$

This can be written in higher level form as

$$\mathbf{A}\underline{x} = \underline{b},$$

where $\mathbf{A}$ denotes the two-by-two matrix above, and $\underline{x}$ and $\underline{b}$ denote the two-by-one vectors. The solution to this equation is then

$$\underline{x} = \mathbf{A}^{-1}\underline{b} = \begin{bmatrix} a & b \\ d & e \end{bmatrix}^{-1} \begin{bmatrix} c \\ f \end{bmatrix}.$$

The general two-by-two matrix inverse is given by

$$\begin{bmatrix} a & b \\ d & e \end{bmatrix}^{-1} = \frac{1}{ae - bd} \begin{bmatrix} e & -b \\ -d & a \end{bmatrix}$$

and the inverse exists whenever $ae - bd$ (which is called the *determinant* of the matrix $\mathbf{A}$) is nonzero. For larger matrices, numerical algorithms are used to invert matrices, such as used by Matlab based on LINPACK [23]. An initial introduction to matrices and linear algebra can be found in [45].

Appendix I

Matlab/Octave Examples

This appendix provides Matlab and Octave examples for various topics covered in this book. The term 'matlab' (uncapitalized) refers to *either* Matlab or Octave [64].

I.1 Complex Numbers in Matlab and Octave

Matlab and Octave have the following primitives for complex numbers:

```
octave:1> help j

j is a built-in constant

 - Built-in Variable: I
 - Built-in Variable: J
 - Built-in Variable: i
 - Built-in Variable: j

A pure imaginary number, defined as `sqrt (-1)'.  The `I' and `J'
forms are true constants, and cannot be modified.  The `i' and `j'
forms are like ordinary variables, and may be used for other
purposes.  However, unlike other variables, they once again assume
their special predefined values if they are cleared *Note Status
of Variables::.

Additional help for built-in functions, operators, and variables
is available in the on-line version of the manual.  Use the command
`help -i <topic>' to search the manual index.

Help and information about Octave is also available on the WWW
```

```
at http://www.octave.org and via the help-octave@bevo.che.wisc.edu
mailing list.

octave:2> sqrt(-1)
ans = 0 + 1i

octave:3> help real
real is a built-in mapper function

 - Mapping Function:  real (Z)
     Return the real part of Z.

See also: imag and conj. ...

octave:4> help imag
imag is a built-in mapper function

 - Mapping Function:  imag (Z)
     Return the imaginary part of Z as a real number.

See also: real and conj. ...

octave:5> help conj
conj is a built-in mapper function

 - Mapping Function:  conj (Z)
     Return the complex conjugate of Z, defined as
     `conj (Z)' = X - IY.

See also: real and imag. ...

octave:6> help abs
abs is a built-in mapper function

 - Mapping Function:  abs (Z)
     Compute the magnitude of Z, defined as
     |Z| = `sqrt (x^2 + y^2)'.

     For example,

          abs (3 + 4i)
          => 5
...
```

```
octave:7> help angle
angle is a built-in mapper function

 - Mapping Function:  angle (Z)
     See arg.
...
```

Note how helpful the "See also" information is in Octave (and similarly in Matlab).

Complex Number Manipulation

Let's run through a few elementary manipulations of complex numbers in Matlab:

```
>> x = 1;
>> y = 2;
>> z = x + j * y

z =
   1 + 2i

>> 1/z

ans =
   0.2 - 0.4i

>> z^2

ans =
  -3 + 4i

>> conj(z)

ans =
   1 - 2i

>> z*conj(z)

ans =
     5

>> abs(z)^2

ans =
     5

>> norm(z)^2

ans =
     5

>> angle(z)

ans =
     1.1071
```

Now let's do polar form:

```
>> r = abs(z)

r =
     2.2361

>> theta = angle(z)

theta =
     1.1071
```

Curiously, e is not defined by default in Matlab (though it is in Octave). It can easily be computed in Matlab as `e=exp(1)`.

Below are some examples involving imaginary exponentials:

```
>> r * exp(j * theta)

ans =
   1 + 2i

>> z

z =
   1 + 2i

>> z/abs(z)

ans =
   0.4472 + 0.8944i

>> exp(j*theta)

ans =
   0.4472 + 0.8944i

>> z/conj(z)

ans =
  -0.6 + 0.8i

>> exp(2*j*theta)

ans =
  -0.6 + 0.8i

>> imag(log(z/abs(z)))

ans =
    1.1071

>> theta

theta =
    1.1071

>>
```

Here are some manipulations involving two complex numbers:

```
>> x1 = 1;
>> x2 = 2;
>> y1 = 3;
>> y2 = 4;
>> z1 = x1 + j * y1;
>> z2 = x2 + j * y2;
>> z1

z1 =
   1 + 3i

>> z2

z2 =
   2 + 4i

>> z1*z2

ans =
 -10 +10i

>> z1/z2

ans =
   0.7 + 0.1i
```

Another thing to note about matlab syntax is that the transpose operator `'` (for vectors and matrices) *conjugates* as well as transposes. Use `.'` to transpose without conjugation:

```
>>x = [1:4]*j

x =
         0 + 1i    0 + 2i    0 + 3i    0 + 4i

>> x'

ans =
         0 - 1i
         0 - 2i
         0 - 3i
         0 - 4i

>> x.'

ans =
         0 + 1i
         0 + 2i
         0 + 3i
         0 + 4i
```

I.2 Factoring Polynomials in Matlab

Let's find all roots of the polynomial

$$p(x) = x^5 + 5x + 7.$$

```
>> % polynomial = array of coefficients in matlab:
>> p = [1 0 0 0 5 7]; %  p(x) = x^5 + 5*x + 7
>> format long;       %  print double-precision
>> roots(p)           %  print out the roots of p(x)

ans =
   1.30051917307206 + 1.10944723819596i
   1.30051917307206 - 1.10944723819596i
  -0.75504792501755 + 1.27501061923774i
  -0.75504792501755 - 1.27501061923774i
  -1.09094249610903
```

I.3 Geometric Signal Theory

This section follows Chapter 5 of the text.

Vector Interpretation of Complex Numbers

Here's how Fig. 5.1 on page 71 may be generated in matlab:

```
>> x = [2 3];                       % coordinates of x
>> origin = [0 0];                  % coordinates of the origin
>> xcoords = [origin(1) x(1)]; % plot() expects coordinates
>> ycoords = [origin(2) x(2)];
>> plot(xcoords,ycoords);           % Draw a line from origin to x
```

Signal Metrics

The *mean* of a signal x stored in a matlab row- or column-vector `x` can be computed in matlab as

```
mu = sum(x)/N
```

or by using the built-in function `mean()`. If `x` is a 2D matrix containing `N` elements, then we need `mu = sum(sum(x))/N` or `mu = mean(mean(x))`, since `sum` computes a sum along "dimension 1" (which is along columns for matrices), and `mean` is implemented in terms of `sum`. For 3D matrices, `mu = mean(mean(mean(x)))`, etc. For a higher dimensional matrices `x`, "flattening" it into a long column-vector `x(:)` is the more concise form:

```
N = prod(size(x))
mu = sum(x(:))/N
```

or

```
mu = x(:).' * ones(N,1)/N
```

The above constructs work whether `x` is a row-vector, column-vector, or matrix, because `x(:)` returns a concatenation of all columns of `x` into one long column-vector. Note the use of `.'` to obtain non-conjugating vector transposition in the second form. `N = prod(size(x))` is the number of elements of `x`. If `x` is a row- or column-vector, then `length(x)` gives the number of elements. For matrices, `length()` returns the greater of the number of rows or columns.[1]

[1] For consistency with `sum` and other functions, it would be better if `length()` returned the number of elements along dimension 1, with the special case of using dimension 2 ("along rows") for row-vectors. However, compatibility with early Matlab dictates the convention used.

Signal Energy and Power

In a similar way, we can compute the *signal energy* $\mathcal{E}_x$ (sum of squared moduli) using any of the following constructs:

```
Ex = x(:)' * x(:)
Ex = sum(conj(x(:)) .* x(:))
Ex = sum(abs(x(:)).^2)
```

The average power (energy per sample) is similarly `Px = Ex/N`. The L^2 norm is similarly `xL2 = sqrt(Ex)` (same result as `xL2 = norm(x)`). The L^1 norm is given by `xL1 = sum(abs(x))` or by `xL1 = norm(x,1)`. The infinity-norm (Chebyshev norm) is computed as `xLInf = max(abs(x))` or `xLInf = norm(x,Inf)`. In general, L^p norm is computed by `norm(x,p)`, with p=2 being the default case.

I.3.1 Inner Product

The inner product $\langle x, y\rangle$ of two column-vectors `x` and `y` (§5.9 on page 80) is conveniently computed in matlab as

```
xdoty = y' * x
```

I.3.2 Vector Cosine

For real vectors `x` and `y` having the same length, we may compute the *vector cosine* by

```
cosxy = y' * x / ( norm(x) * norm(y) );
```

For complex vectors, a good measure of orthogonality is the modulus of the vector cosine:

```
collinearity = abs(y' * x) / ( norm(x) * norm(y) );
```

Thus, when `collinearity` is near 0, the vectors `x` and `y` are substantially orthogonal. When `collinearity` is close to 1, they are nearly collinear.

I.3.3 Projection

As discussed in §5.9.9 on page 86, the orthogonal projection of $y \in \mathbf{C}^N$ onto $x \in \mathbf{C}^N$ is defined by

$$\mathbf{P}_x(y) \triangleq \frac{\langle y, x\rangle}{\|x\|^2} x.$$

In matlab, the projection of the length-`N` column-vector `y` onto the length-`N` column-vector `x` may therefore be computed as follows:

```
yx = (x' * y) * (x' * x)^(-1) * x
```

More generally, a length-N column-vector y can be projected onto the M-dimensional *subspace* spanned by the columns of the N × M matrix X:

```
yX = X * (X' * X)^(-1) * X' * y
```

Orthogonal projection, like any finite-dimensional *linear* operator, can be represented by a matrix. In this case, the $N \times N$ matrix

```
PX = X * (X' * X)^(-1) * X'
```

is called the *projection matrix*.[2] Subspace projection is an example in which the power of matrix linear algebra notation is evident.

Projection Example 1

```
>> X = [[1;2;3],[1;0;1]]
X =

     1     1
     2     0
     3     1

>> PX = X * (X' * X)^(-1) * X'
PX =
```

[2]Going back into "math mode", let's show that the $N \times M$ projection matrix

$$\mathbf{P} = \mathbf{X}(\mathbf{X}^*\mathbf{X})^{-1}\mathbf{X}^* \tag{I.1}$$

yields orthogonal projection onto the column-space of $\mathbf{X}$ (when the $M \times M$ matrix $\mathbf{X}^*\mathbf{X}$ is invertible, and here $\mathbf{X}^*$ denotes the Hermitian (conjugating) transpose of $\mathbf{X}$). That is, if $\hat{y} \triangleq \mathbf{P}y$ is the projection of y onto $\mathbf{X}$, we must have, by definition of orthogonal projection (§5.9.9), that $\hat{y}$ lies in the column-space of $\mathbf{X}$, and that $(y - \hat{y}) \perp \hat{y}$, or $\langle(\mathbf{I}-\mathbf{P})\underline{y}, \mathbf{P}\underline{y}\rangle = 0$ for any $\underline{y} \in \mathbf{C}^N$. We may say that $(\mathbf{I} - \mathbf{P})$ projects onto the *orthogonal complement* of the column-space of $\mathbf{X}$.

That $\hat{y}$ is a linear combination of the columns of $\mathbf{X}$ is immediate because $\mathbf{X}$ is the leftmost term of the definition of $\mathbf{P}$ in Eq. (I.1). To show orthogonality of the "projection error", *i.e.*, that $\langle(\mathbf{I} - \mathbf{P})y, \mathbf{P}y\rangle = 0$ for all $\underline{y} \in \mathbf{C}^N$, note that in matrix notation we must show $\underline{y}^*\mathbf{P}^*(\mathbf{I} - \mathbf{P})\underline{y} = 0$ for all $\underline{y}$, which requires $\mathbf{P}^*(\mathbf{I} - \mathbf{P}) = 0$, or $\mathbf{P}^* = \mathbf{P}^*\mathbf{P}$. Since $\mathbf{P} \triangleq \mathbf{X}(\mathbf{X}^*\mathbf{X})^{-1}\mathbf{X}^*$, it is Hermitian symmetric, so that the orthogonal projection requirement becomes $\mathbf{P} = \mathbf{P}^2$, which is easily verified for $\mathbf{P}$ as defined in Eq. (I.1).

The general property $\mathbf{P}^2 = \mathbf{P}$ defines an *idempotent* square matrix $\mathbf{P}$. Intuitively, it makes sense that a projection should be idempotent, because once a vector is projected onto a particular subspace, projecting again should do nothing. All idempotent matrices are projection matrices, and vice versa. However, only Hermitian (symmetric) idempotent matrices correspond to orthogonal projection [44].

```
   0.66667  -0.33333   0.33333
  -0.33333   0.66667   0.33333
   0.33333   0.33333   0.66667

>> y = [2;4;6]
y =

   2
   4
   6

>> yX = PX * y
yX =

   2.0000
   4.0000
   6.0000
```

Since y in this example already lies in the column-space of X, orthogonal projection onto that space has no effect.

Projection Example 2

Let X and PX be defined as Example 1, but now let

```
>> y = [1;-1;1]
y =

   1
  -1
   1

>> yX = PX * y
yX =

   1.33333
  -0.66667
   0.66667

>> yX' * (y-yX)
ans = -7.0316e-16

>> eps
```

```
ans =   2.2204e-16
```

In the last step above, we verified that the projection yX is orthogonal to the "projection error" y-yX, at least to machine precision. The eps variable holds "machine epsilon" which is the numerical distance between 1.0 and the next representable number in double-precision floating point.

I.3.4 Orthogonal Basis Computation

Matlab and Octave have a function orth() which will compute an orthonormal basis for a space given any set of vectors which span the space. In Matlab, *e.g.*, we have the following help info:

```
>> help orth
 ORTH  Orthogonalization.
       Q = orth(A) is an orthonormal basis for the range of A.
       Q'*Q = I, the columns of Q span the same space as the
       columns of A and the number of columns of Q is the rank
       of A.

       See also QR, NULL.
```

Below is an example of using orth() to orthonormalize a linearly independent basis set for $N = 3$:

```
% Demonstration of the orth() function.
v1 = [1; 2; 3];  % our first basis vector (a column vector)
v2 = [1; -2; 3]; % a second, linearly independent vector
v1' * v2         % show that v1 is not orthogonal to v2

ans =
     6

V = [v1,v2]      % Each column of V is one of our vectors

V =
     1     1
     2    -2
     3     3

W = orth(V)  % Find an orthonormal basis for the same space

W =
    0.2673    0.1690
    0.5345   -0.8452
    0.8018    0.5071
```

```
w1 = W(:,1)  % Break out the returned vectors

w1 =
    0.2673
    0.5345
    0.8018

w2 = W(:,2)

w2 =
    0.1690
   -0.8452
    0.5071

w1' * w2  % Check that w1 is orthogonal to w2

ans =
   2.5723e-17

w1' * w1  % Also check that the new vectors are unit length

ans =
     1

w2' * w2

ans =
     1

W' * W   % faster way to do the above checks

ans =
    1    0
    0    1


% Construct some vector x in the space spanned by v1 and v2:
x = 2 * v1 - 3 * v2

x =
    -1
    10
    -3

% Show that x is also some linear combination of w1 and w2:
c1 = x' * w1      % Coefficient of projection of x onto w1
```

```
c1 =
    2.6726

c2 = x' * w2        % Coefficient of projection of x onto w2

c2 =
  -10.1419

xw = c1 * w1 + c2 * w2  % Can we make x using w1 and w2?

xw =
   -1
   10
   -3

error = x - xw

error = 1.0e-14 *

    0.1332
         0
         0

norm(error)         % typical way to summarize a vector error

ans =
   1.3323e-15

% It works! (to working precision, of course)

% Construct a vector x NOT in the space spanned by v1 and v2:
y = [1; 0; 0];      % Almost anything we guess in 3D will work

%  Try to express y as a linear combination of w1 and w2:
c1 = y' * w1;       % Coefficient of projection of y onto w1
c2 = y' * w2;       % Coefficient of projection of y onto w2
yw = c1 * w1 + c2 * w2  % Can we make y using w1 and w2?
```

```
yw =

    0.1
    0.0
    0.3
```

```
yerror = y - yw

yerror =

    0.9
    0.0
```

```
   -0.3

norm(yerror)
```

```
ans =
    0.9487
```

While the error is not zero, it is the smallest possible error in the least squares sense. That is, `yw` is the optimal least-squares approximation to y in the space spanned by `v1` and `v2` (`w1` and `w2`). In other words, `norm(yerror) is less than or equal to norm(y-yw2)` for any other vector `yw2` made using a linear combination of `v1` and `v2`. In yet other words, we obtain the optimal least squares approximation of `y` (which lives in 3D) in some subspace W (a 2D subspace of 3D spanned by the columns of matrix `W`) by projecting `y` orthogonally onto the subspace W to get `yw` as above.

An important property of the optimal least-squares approximation is that the approximation error is *orthogonal* to the the subspace in which the approximation lies. Let's verify this:

```
W' * yerror   % must be zero to working precision

ans = 1.0e-16 *

   -0.2574
   -0.0119
```

I.4 The DFT

This section gives Matlab examples illustrating the computation of two figures in Chapter 6, and the DFT matrix in Matlab.

I.4.1 DFT Sinusoids for $N = 8$

Below is the Matlab for Fig. 6.2 on page 102:

```
N=8;
fs=1;

n = [0:N-1]; % row
t = [0:0.01:N]; % interpolated
k=fliplr(n)' - N/2;
fk = k*fs/N;
wk = 2*pi*fk;
clf;
for i=1:N
  subplot(N,2,2*i-1);
  plot(t,cos(wk(i)*t))
  axis([0,8,-1,1]);
  hold on;
  plot(n,cos(wk(i)*n),'*')
  if i==1
    title('Real Part');
  end;
  ylabel(sprintf('k=%d',k(i)));
  if i==N
    xlabel('Time (samples)');
  end;
  subplot(N,2,2*i);
  plot(t,sin(wk(i)*t))
  axis([0,8,-1,1]);
  hold on;
  plot(n,sin(wk(i)*n),'*')
  ylabel(sprintf('k=%d',k(i)));
  if i==1
    title('Imaginary Part');
  end;
  if i==N
    xlabel('Time (samples)');
  end;
end
```

I.4.2 DFT Bin Response

Below is the Matlab for Fig. 6.3 on page 107:

```
% Parameters (sampling rate = 1)
N = 16;                  % DFT length
k = N/4;                 % bin where DFT filter is centered
wk = 2*pi*k/N;           % normalized radian center-frequency
wStep = 2*pi/N;
w = [0:wStep:2*pi - wStep]; % DFT frequency grid

interp = 10;
N2 = interp*N; % Denser grid showing "arbitrary" frequencies
w2Step = 2*pi/N2;
w2 = [0:w2Step:2*pi - w2Step]; % Extra dense frequency grid
X = (1 - exp(j*(w2-wk)*N)) ./ (1 - exp(j*(w2-wk)));
X(1+k*interp) = N; % Fix divide-by-zero point (overwrite NaN)

% Plot spectral magnitude
clf;
magX = abs(X);
magXd = magX(1:interp:N2); % DFT frequencies only
subplot(2,1,1);
plot(w2,magX,'-'); hold on; grid;
plot(w,magXd,'*');          % Show DFT sample points
title('DFT Amplitude Response at k=N/4');
xlabel('Normalized Radian Frequency (radians per sample)');
ylabel('Magnitude (Linear)');
text(-1,20,'a)');

% Same thing on a dB scale
magXdb = 20*log10(magX);       % Spectral magnitude in dB
% Since the zeros go to minus infinity, clip at -60 dB:
magXdb = max(magXdb,-60*ones(1,N2));
magXddb = magXdb(1:interp:N2); % DFT frequencies only
subplot(2,1,2);
hold off; plot(w2,magXdb,'-'); hold on; plot(w,magXddb,'*');
xlabel('Normalized Radian Frequency (radians per sample)');
ylabel('Magnitude (dB)'); grid;
text(-1,40,'b)');
print -deps '../eps/dftfilter.eps';
hold off;
```

I.4.3 DFT Matrix

The following example reinforces the discussion of the DFT matrix in §6.12 on page 111. We can simply create the DFT matrix in matlab by taking the DFT of the identity matrix. Then we show that multiplying by the DFT matrix is equivalent to the calling the `fft` function in matlab:

```
>> eye(4)
ans =
     1     0     0     0
     0     1     0     0
     0     0     1     0
     0     0     0     1

>> S4 = fft(eye(4))
ans =
   1         1          1         1
   1         0 - 1i    -1         0 + 1i
   1        -1          1        -1
   1         0 + 1i    -1         0 - 1i

>> S4' * S4              % Show that S4' = inverse DFT (times N=4)
ans =
    4    0    0    0
    0    4    0    0
    0    0    4    0
    0    0    0    4

>> x = [1; 2; 3; 4]
x =
     1
     2
     3
     4
```

```
>> fft(x)
ans =
  10
  -2 + 2i
  -2
  -2 - 2i
```

```
>> S4 * x
ans =
  10
  -2 + 2i
  -2
  -2 - 2i
```

I.5 Spectrogram Computation

This section lists the spectrogram function called in the Matlab code displayed in Fig. 8.11 on page 184.

```
function X = spectrogram(x,nfft,fs,window,noverlap,doplot,dbclip);

%SPECTROGRAM Calculate spectrogram from signal.
% B = SPECTROGRAM(A,NFFT,Fs,WINDOW,NOVERLAP) calculates the
%     spectrogram for the signal in vector A.
%
% NFFT is the FFT size used for each frame of A.  It should be a
% power of 2 for fastest computation of the spectrogram.
%
% Fs is the sampling frequency. Since all processing parameters are
% in units of samples, Fs does not effect the spectrogram itself,
% but it is used for axis scaling in the plot produced when
% SPECTROGRAM is called with no output argument (see below).
%
% WINDOW is the length M window function applied, IN ZERO-PHASE
% FORM, to each frame of A.  M cannot exceed NFFT.  For M<NFFT,
% NFFT-M zeros are inserted in the FFT buffer (for interpolated
% zero-phase processing).  The window should be supplied in CAUSAL
% FORM.
%
% NOVERLAP is the number of samples the sections of A overlap, if
% nonnegative.  If negative, -NOVERLAP is the "hop size", i.e., the
% number of samples to advance successive windows.  (The overlap is
% the window length minus the hop size.)  The hop size is called
% NHOP below.  NOVERLAP must be less than M.
%
% If doplot is nonzero, or if there is no output argument, the
% spectrogram is displayed.
%
% When the spectrogram is displayed, it is "clipped" dbclip dB
% below its maximum magnitude.  The default clipping level is 100
% dB down.
%
% Thus, SPECTROGRAM splits the signal into overlapping segments of
% length M, windows each segment with the length M WINDOW vector, in
% zero-phase form, and forms the columns of B with their
% zero-padded, length NFFT discrete Fourier transforms.
%
% With no output argument B, SPECTROGRAM plots the dB magnitude of
% the spectrogram in the current figure, using
% IMAGESC(T,F,20*log10(ABS(B))), AXIS XY, COLORMAP(JET) so the low
```

```
% frequency content of the first portion of the signal is displayed
% in the lower left corner of the axes.
%
% Each column of B contains an estimate of the short-term,
% time-localized frequency content of the signal A.  Time increases
% linearly across the columns of B, from left to right.  Frequency
% increases linearly down the rows, starting at 0.
%
% If A is a length NX complex signal, B is returned as a complex
% matrix with NFFT rows and
%      k = floor((NX-NOVERLAP)/(length(WINDOW)-NOVERLAP))
%        = floor((NX-NOVERLAP)/NHOP)
% columns.  When A is real, only the NFFT/2+1 rows are needed when
% NFFT even, and the first (NFFT+1)/2 rows are sufficient for
% inversion when NFFT is odd.
%
% See also: Matlab and Octave's SPECGRAM and STFT functions.

if nargin<7, dbclip=100; end
if nargin<6, doplot=0; end
if nargin<5, noverlap=256; end
if nargin<4, window=hamming(512); end
if nargin<3, fs=1; end
if nargin<2, nfft=2048; end

x = x(:); % make sure it's a column

M = length(window);
if length(x)<M, x = [x;zeros(M-length(x),1)]; end;
if (M<2)
  % (Matlab's specgram allows window to be a scalar specifying
  % the length of a Hanning window.)
  error('spectrogram: Expect complete window, not just its length');
end;
Modd = mod(M,2); % 0 if M even, 1 if odd
Mo2 = (M-Modd)/2;
w = window(:); % Make sure it's a column
zp = zeros(nfft-M,1);
wzp = [w(Mo2+1:M);zp;w(1:Mo2)];

if noverlap<0
  nhop = - noverlap;
  noverlap = M-nhop;
else
  nhop = M-noverlap;
end
```

```
nx = length(x);
nframes = 1+floor((nx-noverlap)/nhop);

X = zeros(nfft,nframes);
xoff = 0;
for m=1:nframes-1
  xframe = x(xoff+1:xoff+M); % extract frame of input data
  xoff = xoff + nhop;   % advance in-pointer by hop size
  xzp = [xframe(Mo2+1:M);zp;xframe(1:Mo2)];
  xw = wzp .* xzp;
  X(:,m) = fft(xw);
end

if (nargout==0) | doplot
  t = (0:nframes-1)*nhop/fs;
  f = 0.001*(0:nfft-1)*fs/nfft;
  Xdb = 20*log10(abs(X));
  Xmax = max(max(Xdb));
  % Clip lower limit so nulls don't dominate:
  clipvals = [Xmax-dbclip,Xmax];
  imagesc(t,f,Xdb,clipvals);
  % grid;
  axis('xy');
  colormap(jet);
  xlabel('Time (sec)');
  ylabel('Freq (kHz)');
end
```

Bibliography

[1] M. Abramowitz and I. A. Stegun, eds., *Handbook of Mathematical Functions*, New York: Dover, 1965.

[2] R. Agarwal and C. S. Burrus, "Number theoretic transforms to implement fast digital convolution," *Proceedings of the IEEE*, vol. 63, pp. 550–560, Apr. 1975, also in [41].

[3] J. B. Allen and L. R. Rabiner, "A unified approach to short-time Fourier analysis and synthesis," *Proc. IEEE*, vol. 65, pp. 1558–1564, Nov. 1977.

[4] M. Bosi and R. E. Goldberg, *Introduction to Digital Audio Coding and Standards*, Boston: Kluwer Academic Publishers, 2003.

[5] L. Bosse, "An experimental high fidelity perceptual audio coder," tech. rep., Elec. Engineering Dept., Stanford University (CCRMA), Mar. 1998, Music 420 Project Report, `http://ccrma.stanford.edu/~jos/bosse/`.

[6] K. Brandenburg and M. Bosi, "Overview of MPEG audio: Current and future standards for low-bit-rate audio coding," *Journal of the Audio Engineering Society*, vol. 45, pp. 4–21, Jan./Feb. 1997.

[7] C. S. Burrus, "Notes on the FFT," Mar. 1990.

[8] C. S. Burrus and T. W. Parks, *DFT/FFT and Convolution Algorithms*, New York: John Wiley and Sons, Inc., 1985.

[9] J. P. Campbell Jr., T. E. Tremain, and V. C. Welch, "The proposed federal standard 1016 4800 bps voice coder: CELP," *Speech Technology Magazine*, pp. 58–64, April-May 1990.

[10] D. C. Champeney, *A Handbook of Fourier Theorems*, Cambridge University Press, 1987.

[11] D. G. Childers, ed., *Modern Spectrum Analysis*, New York: IEEE Press, 1978.

[12] J. M. Chowning, "The synthesis of complex audio spectra by means of frequency modulation," *Journal of the Audio Engineering Society*, vol. 21, no. 7, pp. 526–534, 1973, reprinted in [60].

[13] R. V. Churchill, *Complex Variables and Applications*, New York: McGraw-Hill, 1960.

[14] J. Cooley and J. Tukey, "An algorithm for the machine computation of the complex Fourier series," *Mathematics of Computation*, vol. 19, pp. 297–301, Apr. 1965.

[15] J. Dattorro, "The implementation of recursive digital filters for high-fidelity audio," *Journal of the Audio Engineering Society*, vol. 36, pp. 851–878, Nov. 1988, Comments, ibid. (Letters to the Editor), vol. 37, p. 486 (1989 June); Comments, ibid. (Letters to the Editor), vol. 38, pp. 149-151 (1990 Mar.).

[16] J. R. Deller Jr., J. G. Proakis, and J. H. Hansen, *Discrete-Time Processing of Speech Signals*, New York: Macmillan, 1993.

[17] P. A. M. Dirac, *The Principles of Quantum Mechanics, Fourth Edition*, New York: Oxford University Press, 1958–2001.

[18] DSP Committee, ed., *Programs for Digital Signal Processing*, New York: IEEE Press, 1979.

[19] P. Duhamel and M. Vetterli, "Improved Fourier and Hartley algorithms with application to cyclic convolution of real data," *IEEE Transactions on Acoustics, Speech, Signal Processing*, vol. 35, pp. 818–824, June 1987.

[20] P. Duhamel and M. Vetterli, "Fast Fourier transforms: A tutorial review and state of the art," *Signal Processing*, vol. 19, pp. 259–299, Apr. 1990.

[21] M. Frigo and S. G. Johnson, "FFTW: An adaptive software architecture for the FFT," in *Proceedings of the International Conference on Acoustics, Speech, and Signal Processing, Seattle*, vol. 3, (New York), pp. 1381–1384, IEEE Press, 1998, `http://www.fftw.org/`.

[22] B. Gold and C. M. Rader, *Digital Processing of Signals*, New York: McGraw-Hill, 1969.

[23] G. H. Golub and C. F. Van Loan, *Matrix Computations, 2nd Edition*, Baltimore: The Johns Hopkins University Press, 1989.

[24] I. J. Good, "The interaction algorithm and practical Fourier analysis," *Journal of the Royal Statistical Society, Series B*, vol. 20, no. 2, pp. 361–372, 1958, addendum: ibid. 22 (2), 373-375 (1960).

[25] R. M. Gray and L. D. Davisson, *An Introduction to Statistical Signal Processing*, Cambridge University Press, 2003, `http://www-ee.stanford.edu/~gray/sp.pdf`.

[26] J. Gullberg, *Mathematics From the Birth of Numbers*, New York: Norton and Co., 1997, [Qa21.G78 1996] ISBN 0-393-04002-X.

[27] W. Hartmann, *Signals, Sound, and Sensation*, New York: AIP Press, 1997, 647 pp., 221 illustrations, hardcover.

[28] M. Heideman, D. Johnson, and C. S. Burrus, "Gauss and the history of the FFT," *IEEE Signal Processing Magazine*, vol. 1, pp. 14–21, Oct. 1984, also in the *Archive for History of Exact Sciences,* vol. 34, no. 3, pp. 265–277, 1985.

[29] S. G. Johnson and M. Frigo, "A modified split-radix FFT with fewer arithmetic operations," *IEEE Transactions on Signal Processing*, vol. 55, pp. 111–119, Jan. 2007.

[30] T. Kailath, A. H. Sayed, and B. Hassibi, *Linear Estimation*, Englewood Cliffs, NJ: Prentice-Hall, Inc., Apr. 2000.

[31] S. M. Kay, *Fundamentals of Statistical Signal Processing, Volume I: Estimation Theory*, Englewood Cliffs, NJ: Prentice-Hall, Inc., 1993.

[32] S. M. Kay, *Fundamentals of Statistical Signal Processing, Volume II: Detection Theory*, Englewood Cliffs, NJ: Prentice-Hall, Inc., 1998.

[33] D. Kolba and T. Parks, "A prime factor FFT algorithm using high-speed convolution," *IEEE Transactions on Acoustics, Speech, Signal Processing*, vol. 29, pp. 281–294, Aug. 1977, also in [41].

[34] T. I. Laakso, V. Välimäki, M. Karjalainen, and U. K. Laine, "Splitting the Unit Delay—Tools for Fractional Delay Filter Design," *IEEE Signal Processing Magazine*, vol. 13, pp. 30–60, Jan. 1996.

[35] W. R. LePage, *Complex Variables and the Laplace Transform for Engineers*, New York: Dover, 1961.

[36] M. J. Lighthill, *Introduction to Fourier Analysis*, Cambridge University Press, Jan. 1958.

[37] L. Ljung and T. L. Soderstrom, *Theory and Practice of Recursive Identification*, Cambridge, MA: MIT Press, 1983.

[38] J. Makhoul, "Linear prediction: A tutorial review," *Proceedings of the IEEE*, vol. 63, pp. 561–580, Apr. 1975.

[39] J. D. Markel and A. H. Gray, *Linear Prediction of Speech*, New York: Springer Verlag, 1976.

[40] M. V. Mathews, *The Technology of Computer Music*, Cambridge, MA: MIT Press, 1969.

[41] J. H. McClellan and C. M. Rader, *Number Theory in Digital Signal Processing*, Englewood Cliffs, NJ: Prentice-Hall, Inc., 1979.

[42] J. H. McClellan, R. W. Schafer, and M. A. Yoder, *DSP First: A Multimedia Approach*, Englewood Cliffs, NJ: Prentice-Hall, 1998, Tk5102.M388.

[43] B. C. J. Moore, *An Introduction to the Psychology of Hearing*, New York: Academic Press, 1997.

[44] M. S. Moslehian, T. Rowland, and E. W. Weisstein, *Projection Matrix*, From MathWorld–A Wolfram Web Resource, Dec. 2006, `http://mathworld.wolfram.com/ProjectionMatrix.html`.

[45] B. Noble, *Applied Linear Algebra*, Englewood Cliffs, NJ: Prentice-Hall, Inc., 1969.

[46] A. V. Oppenheim and R. W. Schafer, *Digital Signal Processing*, Englewood Cliffs, NJ: Prentice-Hall, Inc., 1975.

[47] D. O'Shaughnessy, *Speech Communication*, Reading MA: Addison-Wesley, 1987.

[48] T. Painter and A. Spanias, "Perceptual coding of digital audio," *Proceedings of the IEEE*, vol. 88, pp. 451–513, Apr. 2000.

[49] A. Papoulis, *Probability, Random Variables, and Stochastic Processes*, New York: McGraw-Hill, 1965.

[50] A. Papoulis, *Signal Analysis*, New York: McGraw-Hill, 1977.

[51] T. W. Parks and C. S. Burrus, *Digital Filter Design*, New York: John Wiley and Sons, Inc., June 1987, contains FORTRAN software listings.

[52] A. D. Pierce, *Acoustics*, American Institute of Physics, for the Acoustical Society of America, 1989, `http://asa.aip.org/publications.html`.

[53] J. R. Pierce, "private communication," 1991.

[54] M. H. Protter and J. Charles B. Morrey, *Modern Mathematical Analysis*, Reading MA: Addison-Wesley, 1964.

[55] L. R. Rabiner and C. M. Rader, eds., *Digital Signal Processing*, New York: IEEE Press, 1972.

[56] L. R. Rabiner and R. W. Schafer, *Digital Processing of Speech Signals*, Englewood Cliffs, NJ: Prentice-Hall, 1978.

[57] L. R. Rabiner, R. W. Schafer, and C. M. Rader, "The chirp z-transform algorithm and its application," *Bell System Technical Journal*, vol. 48, pp. 1249–1292, 1969, also published in IEEE Tr. Audio & Electroacoustics, vol. 17, no. 2, pp. 86-92, 1969.

[58] C. Roads, ed., *The Music Machine*, Cambridge, MA: MIT Press, 1989.

[59] C. Roads, *The Computer Music Tutorial*, Cambridge, MA: MIT Press, 1996.

[60] C. Roads and J. Strawn, eds., *Foundations of Computer Music*, Cambridge, MA: MIT Press, 1985.

[61] W. Rudin, *Principles of Mathematical Analysis*, New York: McGraw-Hill, 1964.

[62] L. L. Sharf, *Statistical Signal Processing, Detection, Estimation, and Time Series Analysis*, Reading MA: Addison-Wesley, 1991.

[63] J. O. Smith, *Techniques for Digital Filter Design and System Identification with Application to the Violin*, PhD thesis, Elec. Engineering Dept., Stanford University (CCRMA), June 1983, CCRMA Technical Report STAN-M-14, `http://ccrma.stanford.edu/STANM/stanms/stanm14/`.

[64] J. O. Smith, *Introduction to Matlab and Octave*, `http://ccrma.stanford.edu/~jos/matlab/`, 2003.

[65] J. O. Smith, *Introduction to Digital Filters with Audio Applications*, `http://ccrma.stanford.edu/~jos/filters/`, Sept. 2007, online book.

[66] J. O. Smith, *Physical Audio Signal Processing*, `http://ccrma.stanford.edu/~jos/pasp/`, Aug. 2007, online book.

[67] J. O. Smith, *Spectral Audio Signal Processing*, `http://ccrma.stanford.edu/~jos/sasp/`, Mar. 2007, online book.

[68] J. O. Smith and J. S. Abel, "Bark and ERB bilinear transforms," *IEEE Transactions on Speech and Audio Processing*, pp. 697–708, Nov. 1999.

[69] J. O. Smith and P. Gossett, "A flexible sampling-rate conversion method," in *Proceedings of the International Conference on Acoustics, Speech, and Signal Processing, San Diego*, vol. 2, (New York), pp. 19.4.1–19.4.2, IEEE Press, Mar. 1984, expanded tutorial and associated free software available at the Digital Audio Resampling Home Page: `http://ccrma.stanford.edu/~jos/resample/`.

[70] H. V. Sorenson, M. T. Heideman, and C. S. Burrus, "On calculating the split-radix FFT," *IEEE Transactions on Acoustics, Speech, Signal Processing*, vol. ASSP-34, pp. 152–156, Feb. 1986.

[71] H. V. Sorenson, D. L. Jones, M. T. Heideman, and C. S. Burrus, "Real-valued fast Fourier transform algorithms," *IEEE Transactions on Acoustics, Speech, Signal Processing*, vol. ASSP-35, pp. 849–863, June 1987.

[72] K. Steiglitz, *A Digital Signal Processing Primer with Applications to Audio and Computer Music*, Reading MA: Addison-Wesley, 1996.

[73] S. S. Stevens and H. Davis, *Hearing: Its Psychology and Physiology*, American Institute of Physics, for the Acoustical Society of America, 1983, copy of original 1938 edition, `http://asa.aip.org/publications.html`.

[74] T. Stilson, *Efficiently Variable Algorithms in Virtual-Analog Music Synthesis—a Root-Locus Perspective*, PhD thesis, Elec. Engineering Dept., Stanford University (CCRMA), June 2006, `http://ccrma.stanford.edu/~stilti/`.

[75] T. Stilson and J. O. Smith, "Alias-free synthesis of classic analog waveforms," in *Proceedings of the 1996 International Computer Music Conference, Hong Kong*, Computer Music Association, 1996, `http://ccrma.stanford.edu/~stilti/`.

[76] R. D. Strum and D. E. Kirk, *First Principles of Discrete Systems and Digital Signal Processing*, Reading MA: Addison-Wesley, 1988.

[77] L. H. Thomas, "Using a computer to solve problems in physics," in *Applications of Digital Computers*, Boston: Ginn, 1963.

[78] M. Unser, "Splines: A perfect fit for signal and image processing," *IEEE Signal Processing Magazine*, vol. 16, pp. 22–38, Nov. 1999.

[79] V. Välimäki, *Discrete-Time Modeling of Acoustic Tubes Using Fractional Delay Filters*, PhD thesis, Report no. 37, Helsinki University of Technology, Faculty of Electrical Engineering, Laboratory of Acoustic and Audio Signal Processing, Espoo, Finland, Dec. 1995, `http://www.acoustics.hut.fi/~vpv/publications/-vesa_phd.html`.

[80] (Various), *Prime-Factor FFT Algorithm*, `http://www.wikipedia.org/wiki/Prime-factor_FFT_algorithm`, 2003.

[81] G. N. Watson, *A Treatise on the Theory of Bessel Functions*, Cambridge University Press, 2 ed., 1944.

[82] P. D. Welch, "The use of fast Fourier transforms for the estimation of power spectra: A method based on time averaging over short modified periodograms," *IEEE Transactions on Audio and Electroacoustics*, vol. 15, pp. 70–73, 1967, reprinted in [11] and [55].

[83] U. Zölzer, ed., *DAFX—Digital Audio Effects*, West Sussex, England: John Wiley and Sons, LTD, 2002, `http://www.dafx.de/`.

[84] E. Zwicker and H. Fastl, *Psychoacoustics: Facts and Models*, Berlin: Springer Verlag, 1999, second updated edition, 80pp., CD-ROM/softcover.

Index

13519364R00177

Made in the
USA
Monee, IL